SpringerBriefs in Space Life Sciences

More information about this series at http://www.springer.com/series/11849

Stefan Schneider
Editor

Exercise in Space

A Holistic Approach for the Benefit of Human Health on Earth

Editor
Stefan Schneider
German Sport University Cologne
Cologne, Germany

ISSN 2196-5560 ISSN 2196-5579 (electronic)
SpringerBriefs in Space Life Sciences
ISBN 978-3-319-29569-5 ISBN 978-3-319-29571-8 (eBook)
DOI 10.1007/978-3-319-29571-8

Library of Congress Control Number: 2016938910

Printed on acid-free paper

This Springer imprint is published by Springer Nature
The registered company is Springer International Publishing AG Switzerland

Preface to the Series

The extraordinary conditions in space, especially microgravity, are utilized today not only for research in the physical and materials sciences—they especially provide a unique tool for research in various areas of the life sciences. The major goal of this research is to uncover the role of gravity with regard to the origin, evolution, and future of life and to the development and orientation of organisms from single cells and protists up to humans. This research only became possible with the advent of manned spaceflight some 50 years ago. With the first experiment having been conducted onboard Apollo 16, the German Space Life Sciences Program celebrated its 40th anniversary in 2012—a fitting occasion for Springer and the DLR (German Aerospace Center) to take stock of the space life sciences achievements made so far.

The DLR is the Federal Republic of Germany's National Aeronautics and Space Research Center. Its extensive research and development activities in aeronautics, space, energy, transport, and security are integrated into national and international cooperative ventures. In addition to its own research, as Germany's space agency the DLR has been charged by the federal government with the task of planning and implementing the German space program. Within the current space program, approved by the German government in November 2010, the overall goal for the life sciences section is to gain scientific knowledge and to reveal new application potentials by means of research under space conditions, especially by utilizing the microgravity environment of the International Space Station ISS.

With regard to the program's implementation, the DLR Space Administration provides the infrastructure and flight opportunities required, contracts the German space industry for the development of innovative research facilities, and provides the necessary research funding for the scientific teams at universities and other research institutes. While so-called small flight opportunities like the drop tower in Bremen, sounding rockets, and parabolic airplane flights are made available within the national program, research on the International Space Station ISS is implemented in the framework of Germany's participation in the ESA Microgravity Program or through bilateral cooperations with other space agencies. Free flyers such as BION or FOTON satellites are used in cooperation with Russia. The recently started utilization of Chinese spacecraft like Shenzhou has further expanded Germany's spectrum of flight opportunities, and discussions about future cooperation on the planned Chinese Space Station are currently under way.

From the very beginning in the 1970s, Germany has been the driving force for human spaceflight as well as for related research in the life and physical sciences in Europe. It was Germany that initiated the development of Spacelab as the European contribution to the American Space Shuttle System, complemented by setting up a sound national program. And today Germany continues to be the major European contributor to the ESA programs for the ISS and its scientific utilization.

For our series, we have approached leading scientists first and foremost in Germany, but also—since science and research are international and cooperative endeavors—in other countries to provide us with their views and their summaries of the accomplishments in the various fields of space life sciences research. By presenting the current SpringerBriefs on muscle and bone physiology, we start the series with an area that is currently attracting much attention—due in no small part to health problems such as muscle atrophy and osteoporosis in our modern ageing society. Overall, it is interesting to note that the psychophysiological changes that astronauts experience during their spaceflights closely resemble those of ageing people on Earth but progress at a much faster rate. Circulatory and vestibular disorders set in immediately, muscles and bones degenerate within weeks or months, and even the immune system is impaired. Thus, the ageing process as well as certain diseases can be studied at an accelerated pace, yielding valuable insights for the benefit of people on Earth as well. Luckily for the astronauts: these problems slowly disappear after their return to Earth, so that their recovery processes can also be investigated, yielding additional valuable information.

Booklets on nutrition and metabolism, on the immune system, on vestibular and neuroscience, on the cardiovascular and respiratory system, and on psychophysiological human performance will follow. This separation of human physiology and space medicine into the various research areas follows a classical division. It will certainly become evident, however, that space medicine research pursues a highly integrative approach, offering an example that should also be followed in terrestrial research. The series will eventually be rounded out by booklets on gravitational and radiation biology.

We are convinced that this series, starting with its first booklet on muscle and bone physiology in space, will find interested readers and will contribute to the goal of convincing the general public that research in space, especially in the life sciences, has been and will continue to be of concrete benefit to people on Earth.

Bonn, Germany — Günter Ruyters
Bonn, Germany — Markus Braun
July, 2014

DLR Space Administration in Bonn-Oberkassel (DLR)

The International Space Station ISS; photo taken by an astronaut from the space shuttle Discovery, March 7, 2011 (NASA)

Extravehicular activity (EVA) of the German ESA astronaut Hans Schlegel working on the European Columbus lab of ISS, February 13, 2008 (NASA)

Foreword

After the publication of the first two volumes of this series on musculoskeletal problems in December 2014 and on nutrition physiology and metabolism in spring 2015, respectively, we now present the next volume on the topic of "Exercise in Space—a holistic approach for the benefit of human health on Earth". As is pointed out by the authors, human habitats in space such as the International Space Station ISS represent very special cases of extreme living conditions. However, similar isolated and stressful conditions—except microgravity, of course—prevail also on Earth, e.g., in Antarctic stations, submarines, and other extreme environments. All these special conditions challenge not only the physiological systems of the human body but also the psychological system. Exercise has positive effects on both. In this sense, development of new exercise countermeasures for astronauts together with accompanying scientific research is also beneficial for physiological and mental health care on Earth especially for the ageing society of the industrialized countries.

In a highly interdisciplinary and integrative approach, the authors describe in five chapters the challenges of physiological and psychological systems provoked by extreme conditions such as the ones prevailing in space. Since exercise requires certain motor skills, the booklet starts with a description of the deficits that motor skills experience in microgravity and advertises mental training as a countermeasure. In the second chapter, the adaptation of cartilage to immobilization—provoked by bedrest studies, in bedridden patients, or in spaceflight—is investigated. This is a highly neglected topic also in medicine on Earth, although healthy cartilage is essential for the unconfined functioning of the musculoskeletal system. Again, human spaceflight provides a unique opportunity to study these changes that normally occur over months and years on Earth so to say in time lapse within a few weeks in microgravity. The following chapter concludes the pure physiological aspects of the booklet by describing the regulation of the cardio-respiratory system, a system critical for the body's work capacity. Especially, the effects of microgravity on aerobic capacity in the context of exercise are in the center of interest.

The authors of the fourth chapter advertise exercise not only for the physiological fitness but also for social well-being and mental health. This line of thought is expanded by the authors of the last chapter, describing the positive neurocognitive and neuro-affective effects of exercise on human beings. Since spaceflight leads to physiological deconditioning and to mental impairments at the same time, future countermeasures and exercise regimes need to take both aspects into consideration.

All in all, the authors succeed in achieving an interesting balance between physiological and psychological aspects as well as between exercise and countermeasures developed for astronauts in space and their applications on Earth for the benefit of human health—a successful holistic approach.

Bonn, Germany
December 2015

Günter Ruyters

Preface

Living in weightlessness is living under extreme conditions. But it is not only the International Space Station (ISS) that provides these unique conditions. Whereas on the ISS there is an additional stressor, i.e., the absence of gravity, similar living conditions are present in Arctic/Antarctic stations (e.g., the new German Neumayer III station), drilling platforms, and just lately the miners being accidentally isolated in Chile.

Within the recent years, the German Sport University has developed a leading role describing the degenerative effects of living in weightlessness and extreme environments. Since the first German space lab mission (D1), scientists of the German Sport University have been involved into the planning and execution of research programs in order to help maintain the health of human beings living in weightlessness and extreme conditions. In the first decades, this research aimed to understand the impact of weightlessness on different physiological systems. But with ongoing mission duration, the end of the space shuttle era and the beginning of MIR and ISS era, the interest also into psychophysiological problems arose.

While the absence of gravity primarily challenges our physiological system (cardiovascular, musculoskeletal, cartilage, immune system) living in an isolated environment is accompanied by social and mental restraints. As degenerative processes are extremely accelerated in those extreme environments, all these stressors sum up in a way that could be described as a time lapse of everyday life in a more and more sedentary society (Fig. 1). When leaving earth gravity, there is no more physical load on an astronaut's musculoskeletal system for up to 6 months. This is what happens—of course in a more moderate and decelerated form—to the ageing human being. During a mission, astronauts need to arrange with a number of individuals as they are forced (voluntarily or not) to live in an isolated and restricted environment. In everyday life, we are enclosed in different social groups and we need to develop strategies to cope with these groups and their individual members. Both, living in extreme conditions and living in the modern society, is accompanied by different stressors, which could be classified as social (work-life balance), mental (high workload) and physical (sedentary lifestyle) stressor. Such stressors have shown to impact on different psychological as well as physiological systems and might therefore negatively affect health as defined by World Health Organization (WHO) in 1946 "[...] a state of complete physical, mental and social well-being [...]."

Throughout the last decades, exercise has gained an important role when it comes to health prevention. Today astronauts and cosmonauts are exercising an average of up to 2 h per day to counteract the loss of muscle and bone mass, to improve cardiovascular fitness, and to qualify for a quicker rehabilitation phase post-flight. Today exercise is perceived as a unique and holistic tool to counteract physiological deconditioning. But exercise might also be applicable when it comes to the prevention of psychological diseases like stress and burnout, and even a positive effect of a regular exercise regimen on the probability to develop Alzheimer's disease and other neurodegenerative diseases has been shown. Exercise has been proven to have a positive impact on social, mental, and physiological stressors and to positively affect the described psychological and physiological subsystems (Fig. 1).

Besides the fact that we are dealing with the human being living under extreme conditions, we need to be aware of the fact that space science research serves the human being on earth. As the degenerative processes in weightlessness and/or extreme conditions are tremendously accelerated and effects are potentiated, research in this area helps us to understand the underlying mechanism and therefore provide benefit also for the general population. Living in space under extreme conditions provides us with a time lapse of deconditioning of the human physiological and psychological systems and offers the opportunity to develop, foresee, and counteract the degeneration of the human physiology in our civilized world and to stress the importance of exercise on physical and mental health.

The aim of this book is to bring together researchers from different disciplines and to let them share their knowledge and their experience, gained over decades, on how such an extreme environment like space or analogue environment affects different subsystems of the human being and how exercise can be used to counteract negative effects of immobilization. Although coming from different backgrounds, from natural as well as social and cognitive sciences, all authors agree on the positive effects of exercise. Finally, and disengaged from the individual chapters, this book proves that an interdisciplinary, integrative approach is necessary in order to achieve a deeper understanding of the optimization, preservation, and the improvement of health aspects under the extreme conditions in space.

Cologne, Germany Stefan Schneider

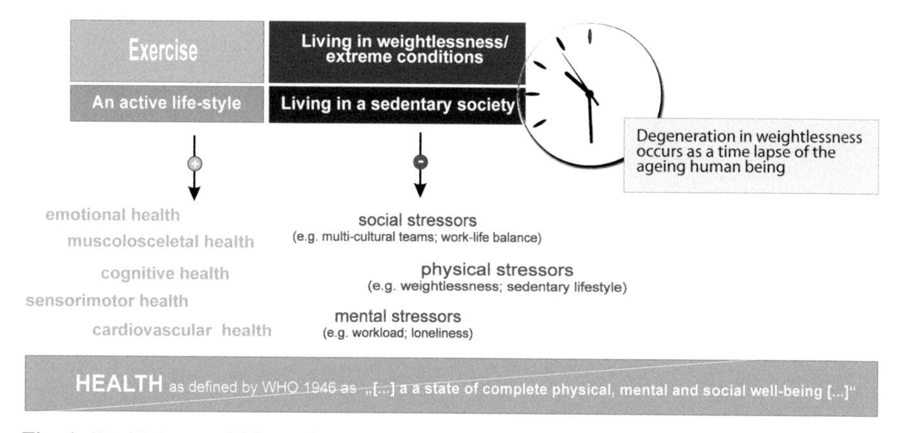

Fig. 1 Health is a multidimensional state summing up physical, mental, and social aspects of well-being. Whereas living in extreme conditions as an analogue for a modern, sedentary lifestyle negatively affects subdimensions of health, exercise and/or an active lifestyle is able to positively affect social, physical, and mental health

Contents

Chapter 1
Motor Skills

Otmar Bock

Abstract Physical exercise is an important countermeasure during Space missions. Since exercising requires motor skills, this chapter reviews the known deficits of motor skills in weightlessness and points out how preflight mental practice might improve motor skills in weightlessness and thus increase the efficiency of onboard workouts.

Keywords Manual skills • Dexterity • Spaceflight • Motor imagery • Mental practice

1.1 Introduction

Physical exercise is an important daily routine aboard the International Space Station (ISS): each crew member spends about 2 h per day exercising, to counteract the deconditioning of gravity-dependent biological systems, such as the musculo-skeletal and the cardiovascular systems (White and Averner 2001). In addition, regular workout is a countermeasure to prevent negative changes of crew members' mood (Schneider et al. 2010), which otherwise might arise due to the prevalence of stressors such as confinement, danger, lack of privacy, and changes of the circadian rhythm. Physical exercise therefore keeps astronauts fit and ready for extravehicular activities, planetary exploration, and return to Earth.

The ISS offers a variety of options for exercising, from simple bungee cords to cycle ergometers, treadmills, and an advanced resistive exercise device (see Fig. 1.1). Using this equipment requires a certain degree of motor skills, as it does on Earth, but skills are known to deteriorate in weightlessness. A workout aboard the Space station may therefore not be as effective as it is on Earth or may even precipitate accidents. This chapter reviews the deficits of motor skills in weightlessness, points out the consequences for physical exercise, and suggests possible remedies. In keeping with established taxonomy (Burton and Miller 1998),

O. Bock (✉)
Institute for Physiology and Anatomy, German Sport University, Köln, Germany
e-mail: Bock@dshs-koeln.de

S. Schneider (ed.), *Exercise in Space*, SpringerBriefs in Space Life Sciences,
DOI 10.1007/978-3-319-29571-8_1

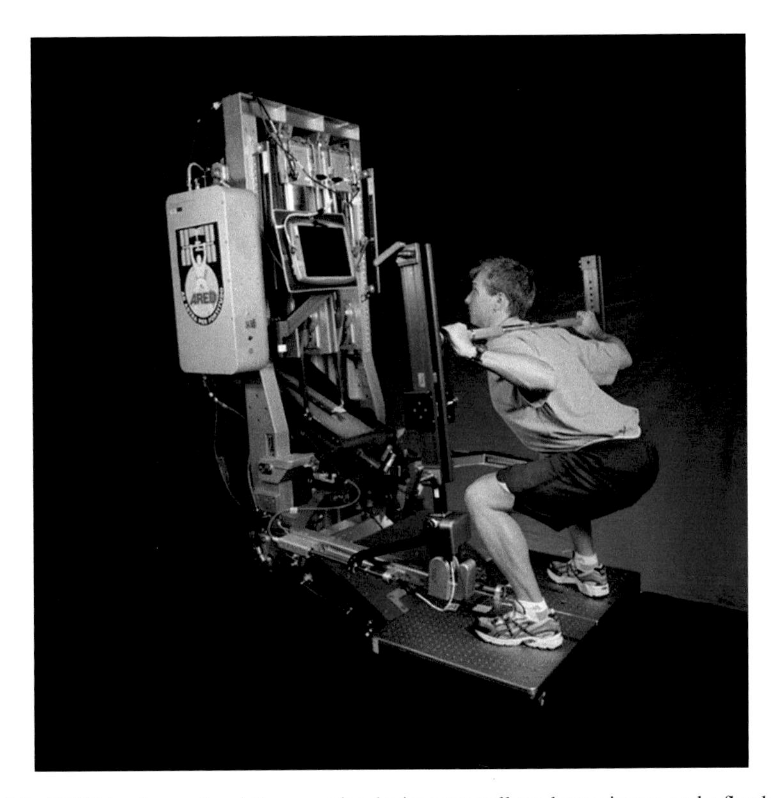

Fig. 1.1 NASA's advanced resistive exercise device uses pulleys, bars, pistons, and a flywheel to provide adjustable resistive and inertial forces; it can be configured for a wide range of exercises such as shoulder presses, heel rises, and free weight workouts (*Source*: http://spaceflight.nasa.gov/gallery/images/shuttle/sts-126/hires/jsc2005-00104.jpg)

the postural, locomotive, and manipulative components of motor skills are addressed separately.

It should be noted, however, that our knowledge about motor skills in weightlessness is severely limited, for several reasons. First, only few studies about motor skills during Space missions are available in literature, some of them lacking scientific rigor. Second, some of the available studies dealt with perceptual phenomena and may be of little relevance for motor control: it is well established that perception and action may obey different functional principles (Aglioti et al. 1995; Bridgeman et al. 1989) and use different neuroanatomical pathways (Mishkin et al. 1983; Goodale et al. 1994), and perceptual deficits might therefore be larger, smaller, or qualitatively different when compared to motor deficits. Third, the majority of studies used typical laboratory tasks in which subjects produced externally triggered, repetitive, simple movements without a behavioral context; in contrast, realistic skills such as those involved in exercising are volitionally initiated, varied, and part of a complex and ecologically valid activity. It is known that such realistic tasks differ from standard laboratory tasks with respect to movement kinematics and dynamics (Bock and Züll 2013) and that the differences vary in dependence on subjects' cognitive abilities (Steinberg and Bock 2012) and exposure to weightlessness (Steinberg and Bock 2013). Inferences about exercising in weightlessness, derived from available microgravity data, must therefore be regarded with caution.

1.2 Posture in Weightlessness

Postural control maintains the stability of our body with respect to the environment. This is achieved through reflexes involving mainly the vestibular, proprioceptive, tactile, and visual system and through volitional adjustments based on our internal representation of Space. Exposure to weightlessness has a profound effect both on the input and on the output side of postural control. The information provided by input signals is affected in that the vestibular system no longer registers head orientation with respect to the upright, articular proprioceptors and plantar touch receptors no longer provide pressure cues about the upright, and visual information can be misleading, e.g., when a fellow crew member drifts by in an unusual body orientation. The output requirements change as well; as an example, forward arm movements on Earth require postural responses that prevent falling forward, but forward arm movements of a free-floating astronaut require responses that prevent backward displacements of the whole body.

Available literature documents that the neutral body posture in weightlessness differs from the erect posture on Earth. Astronauts assume a characteristic flexion of trunk and legs (Thornton et al. 1977), due to a redistribution of muscular tone between flexors and extensors (Clément et al. 1984). Postural reflexes in response to perturbations such as forward arm movements adapt almost instantly to weightlessness (Clément et al. 1985; Clément and Lestienne 1988; Baroni et al. 1999). This has been documented under conditions where the feet are anchored to the spacecraft floor and full vision of the spacecraft interior is available; however, astronauts are rarely fixed to the floor (Tafforin et al. 1989), and little is known about the adequacy of postural reflexes when free floating.

Due to the absence of ground reaction forces, astronauts must find other means to stabilize their own body while interacting with objects in the spacecraft; if not, forces applied to the object would move their body in the opposite direction, as per Newton's third law. A stable posture can be achieved in weightlessness by fastening the feet in footloops, grasping a handlebar, or wedging the body between pieces of equipment (Tafforin et al. 1989). All these are voluntary actions rather than postural reflexes and therefore are cognitively demanding. Furthermore, such actions are planned and executed in reference to an internal representation of Space (Wolpert et al. 1995) which, however, is degraded by weightlessness in multiple ways. The pull of gravity as a fundamental reference for spatial orientation is no longer available (Lackner and DiZio 1993; Glasauer and Mittelstaedt 1998), which induces visuospatial illusions and spatial memory problems (Lackner and DiZio 1993) that persist even after several months in Space (Oman et al. 1986; Kornilova 1997). Weightless persons may lose all sense of orientation when closing their eyes

(Lackner and Graybiel 1979; Glasauer and Mittelstaedt 1992) or entering a less familiar spacecraft module (Oman 2007). The sense of verticality becomes ambiguous: astronauts align their subjective vertical with visual references such as spacecraft architecture or a fellow crew member's body and also with intrinsic references such as the own long body axis, depending on individual preference, environmental cues, and task assignment (Harm and Parker 1993; Oman et al. 1986). Summing up, spatial orientation is less reliable in weightlessness compared to Earth, which degrades the ability of astronauts to volitionally stabilize their body.

1.3 Locomotion in Weightlessness

When we move from one place to another on Earth, our motor system takes advantage of ground reaction forces to accelerate the body forward with one foot and to decelerate it again with the other foot. The kinematics of this behavior is quite stereotyped and is controlled by autonomous neural pattern generators situated in the spinal cord. In weightlessness, however, ground reaction forces are absent. Astronauts therefore move through the Space station in dramatically different ways: they gently push off a surface and then slowly move in free float, or they grasp objects with their hands and then pull their body toward or push it away from those objects. Locomotion in weightlessness therefore is not automated as on Earth but rather is a cognitively demanding motor act.

To arrive at the desired place, we must plan and implement a suitable movement path, taking into account architecture and obstacles. On Earth, we often navigate with the help of a cognitive map; alternatively, we may follow a previously learned route or move from one landmark to the next. The use of a cognitive map is likely to be less effective in weightlessness since spatial orientation is degraded (see above); indeed, anecdotal introspective reports suggest that astronauts rely quite heavily on route and landmark navigation (Oman 2007).

A mental representation of the own movement path is easy to establish when the path involves only fore-aft translations, left-right translations, and rotations about the long body axis (Vidal et al. 2004); these are the three axes normally involved during locomotion on Earth. It therefore is not surprising that astronauts tend to limit locomotion to these three axes where possible (Tafforin and Campan 1994). However, their daily work often requires them to move along unusual axes, e.g., during transit between ISS modules or during work in tight and cluttered quarters; such situations therefore represent a challenge for locomotor planning.

When astronauts use a treadmill for working out, they tie their body down with a subject loading system (bungee cords or cables) to produce ground reaction forces. In this situation, gait kinematics is quite similar to that on Earth (DeWitt et al. 2014), which probably reflects the robustness of spinal pattern generators. The quality of locomotion has not yet been evaluated experimentally, although it is frequently monitored operationally. This is to ensure crew member safety and health adhering to good exercise form, thus preventing injury, especially in the critical phases: early in the mission, when the astronaut is required to adapt to exercise in weightlessness, and shortly before return to earth, when high exercise loads are used. These observations of video downlinks indicate that the speed and quality of body movements are reduced, especially in the first weeks onboard the ISS, but it improves significantly over the course of the mission (N. Petersen, ESA-EAC, personal communication).

Astronauts have consistently reported that navigating through the Space station can be problematic, particularly when they moved between modules whose architectural verticals were not aligned; their reports included instances of disorientation and the use of landmarks instead of cognitive maps (Oman 2007). Several effects of weightlessness on navigation have been documented even at the neurophysiological level (Cheron et al. 2014).

1.4 Manipulation in Weightlessness

Once we are in a stable position near an object of interest, we can manipulate it by extending our hand toward it (transport component), forming our hand to approximate the object's size and shape (grasp component), and exerting forces on it (grip force component). This requires knowledge about the spatial characteristics of object and hand, accurate feedback about the momentarily produced hand forces, and the ability to plan and produce motor commands that actually bring the hand into the desired location with the desired shape.

The relative location of two objects in the frontal plane is judged accurately in weightlessness (Friederici and Levelt 1990; Lipshits and McIntyre 1999), but the egocentric distance (Clément et al. 2008), vertical length (Clément et al. 2012), and depth of objects (Villard et al. 2005) are misjudged. The limb position sense is degraded (Lackner and DiZio 1993; Schmitt and Reid 1985), possibly because of reduced muscle tone (Clément et al. 1985), and the limb position achieved by a given motor command is different than on Earth since the limb is not pulled down by gravity. Furthermore, due to the absence of ground reaction forces, an attempt to move an object may result in a movement of the crew member rather than of the object.

The direction of gravity serves as a fundamental reference for spatial orientation on Earth (Clément and Ngo-Anh 2013), but it is no longer available in weightlessness, which encumbers the planning of limb movements. As an example, NASA's integration standards stipulate that switches aboard the ISS shall turn a process on when they are flipped up and turn it off when they are flipped down (National Aeronautics and Space Administration 2010); however, the meaning of "up" and "down" is ambiguous in weightlessness (see above).

Pointing movements of the arm were found to be less accurate during spaceflight than on Earth (Bock et al. 2010; Watt 1997) and/or slower (Bock et al. 2001; Berger et al. 1997) and/or more resource demanding (Bock et al. 2001; Manzey et al. 1995), particularly with regard to motor programming resources (Bock et al. 2010). At the neurophysiological level, premovement EEG differs from that on ground (Cheron et al. 2014). Grasping movements were found to slow down in weightlessness (Bock et al. 2003), and the ratio of grip-to-load force magnitudes changed, while the temporal coupling of both components didn't change (Nowak et al. 2001). The speed and accuracy of tracking movements were found to change little in weightlessness (Bock et al. 2003), unless the tracking task was highly demanding (Manzey et al. 1995). Thus summing up, the influence of weightlessness on motor skills was inconsistent, in that speed and/or accuracy and/or resource demand could be affected. Figure 1.2 provides an example: the same subjects tested in the same session with the same device exhibited a reduced movement speed for pointing but not for tracking movements in weightlessness. Likewise, surgical skills were found to be less accurate in some studies (Panait et al. 2006) and (Rafiq et al. 2005) and accurate but slower in another study (Campbell et al. 2005). The inconsistent effects of weightlessness on motor skills led to the formulation of a three-factor hypothesis according to which motor performance reflects a trade-off between speed, accuracy (Fitts 1954), and cognitive expenditure (Navon and Gopher 1979): when faced with the challenge of weightlessness, our motor system makes a strategic decision which factor(s) to protect at the expense of which other factor(s) (Bock et al. 2003). As an example, daily routines such as teeth brushing may be executed at lower speed than on Earth but with unchanged accuracy and cognitive involvement, while critical activities such as spacecraft landing maneuvers may call for unchanged speed and accuracy at the expense of a higher

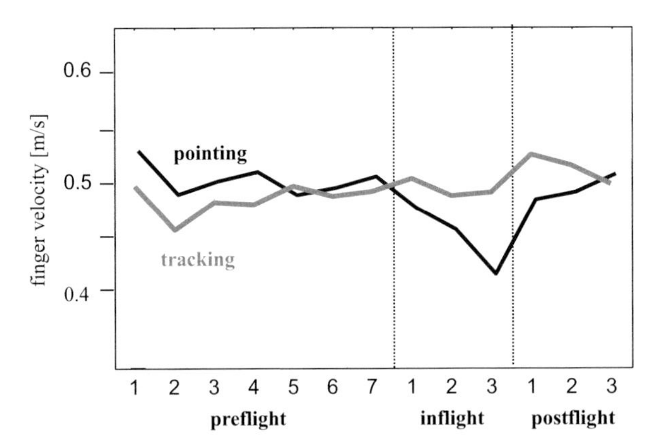

Fig. 1.2 Velocity of skilled finger movements in several experimental sessions before, during, and after a 16-day Space shuttle mission. The velocity of pointing movements (*black*) decreases continuously during the mission and recovers quickly after landing; the decrease therefore probably reflects a compensatory strategy. In contrast, the velocity of pursuit tracking (*gray*) doesn't decrease. Our interpretation is that in pointing, where a lower speed has no overt consequences, participants slowed down to overcome the challenges of weightlessness. In tracking, where a lower speed would immediately produce a pursuit error, participants rather invested extra cognitive resources to overcome these challenges (*Data source*: Bock et al. 2003)

cognitive involvement than on Earth. Such strategic decisions will depend on task constraints and on individual preferences.

It is still a matter of debate exactly why motor skills are challenged by weightlessness. Contributing factors could be the direct effects of weightlessness on sensory organs and on the limbs (Bock et al. 1992), the ongoing adaptive restructuring of the sensorimotor system (Bock 1998), and the presence of high stress levels because of time pressure, lack of privacy, imminent danger, etc. (Fowler et al. 2000).

1.5 Motor Skills as a Constraint for Physical Exercise in Weightlessness

To exercise on the Space station, equipment must be unstowed, assembled, used, dissembled, and restowed, all activities that require a good deal of motor skills. Postural control is needed to keep the body stable while interacting with the equipment, thus avoiding disorientation and undesired movements of the crew member rather than of equipment parts. Locomotion is needed to navigate through the station while transporting the own body and equipment parts back and forth between lockers and the setup area, avoiding obstacles along the way. Manipulation is needed to remove from the locker, unwrap, unfold, connect, adjust, and tie down equipment during setup, to apply forces and displacements to the equipment in the course of exercising, and to untie, disconnect, fold, wrap, and insert it into the locker after use. Since all three components of motor skills are adversely affected by weightlessness (Fig. 1.3), physical exercise might be less efficient during a mission than it is on Earth. Activities before, during, and after the workout may take longer and/or be less accurate, the likelihood of mistakes may increase, and, as a consequence, the musculoskeletal load distribution could be less and then optimal and accidents could occur.

The detrimental effects of weightlessness on motor skills are not constant throughout a Space mission but rather decrease with time and practice. Adaptation to weightlessness follows multiple time constants: the coupling of grip and load forces normalizes almost instantly (Hermsdörfer et al. 2000), simple and well-practiced skills normalize within hours to days (Lackner and DiZio 1996), while complex and novel skills don't fully recover even after several months in Space (Bock et al. 2010; Manzey et al. 1995). These improvements are cognitively demanding, and aftereffects upon return to Earth are limited (Bock et al. 2001, 2010; Manzey et al. 1995). Generally accepted reasoning (Redding and Wallace 1996; McNay and Willingham 1998) therefore suggests that the improvements are largely based on cognitive workaround strategies rather than on "true" recalibration of the sensorimotor system (Bloomberg and Bock 2012). The switch from navigation by cognitive maps to navigation by routes and landmarks (see above) is one example for such cognitive workaround strategies.

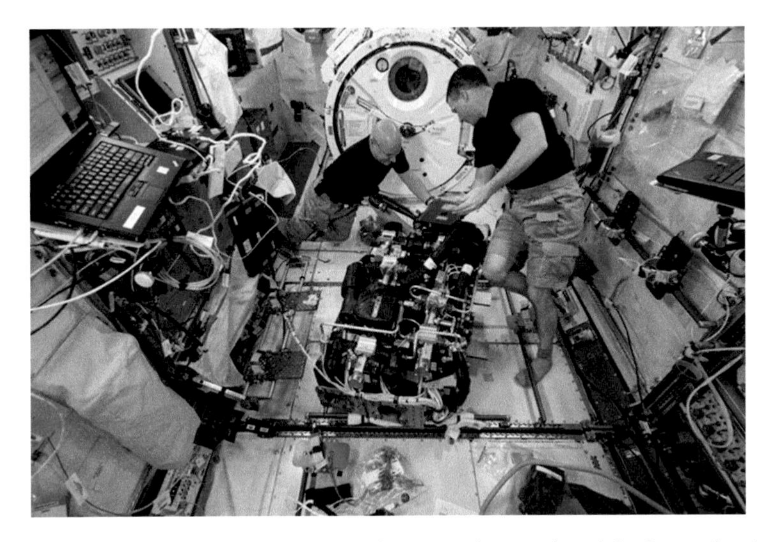

Fig. 1.3 An exemplary view of astronauts setting up equipment aboard the International Space Station. Posture, locomotion, and manipulation must cope with challenges that are absent on Earth; therefore, motor skills are less automated and more cognitively demanding. The direction "down" becomes ambiguous: it could be defined as "toward the floor," "toward my own feet," or "toward the feet of my teammate," and the definition could change repeatedly (*Source*: http://www.nasa.gov/mission_pages/station/images/index.html?id=358587)

Thus summing up, physical exercise relies on motor skills which are degraded early during a mission and partly recover later on at the expense of an increased cognitive load. As the extra cognitive load has adverse biological and psychological consequences (Robert and Hockey 1997), it seems advisable to schedule some of the required cognitive processing for preflight training and thus to reduce the cognitive burden aboard the Space station.

1.6 Preflight Training of Physical Exercise

Astronauts are thoroughly trained on all tasks scheduled on their upcoming mission, including physical exercise. This ground-based training allows astronauts to become familiar with the equipment, with the procedures for its unstowage, setup, use, and stowage, as well as with the limb and body movements needed to carry out those procedures. However, training on Earth proceeds in the presence of ground reaction forces, of gravitational pull on the arm, and of normal sensory inputs; astronauts therefore acquire motor skills which are not necessarily adequate in weightlessness. As a consequence, it might be necessary to unlearn those skills—or components thereof—and replace them by new ones. This will naturally occur once crew members are onboard, but it will require time and cognitive resources (see above). Ground-based preflight training therefore engenders both benefits and costs.

Some training is scheduled during water immersion, where ground reaction forces and the downward pull on the limbs are largely eliminated by buoyancy. This training is presently limited to extravehicular activities and therefore doesn't cover physical exercise. However, it is quite conceivable that skills acquired underwater will transfer from trained to untrained tasks such as physical exercise. Unfortunately, however, immersion differs from weightlessness in multiple ways. Gravity continues to stimulate the vestibular organs, viscous forces oppose all body movements, the visual environment is blurred and magnified, and cognitive performance is degraded due to increased ambient pressure (Dalecki et al. 2012). Thus again, astronauts may have to unlearn some of the motor skills acquired underwater once they are weightless and replace them by new ones.

Some training, again excluding physical exercise, is completed during parabolic flight campaigns where participants can experience true free floating. Unfortunately, however, parabolic flights are expensive, rare, and offer only brief periods of weightlessness. The benefits of parabolic flight training are therefore necessarily limited.

We have recently proposed another approach for preflight training of motor skills (Bock et al. 2015). This approach, mental practice, has been used successfully in the past to train professional athletes. It consists of a training regime in which participants mentally enact well-defined body movements from a first-person perspective, without any overt motor output (Decety and Grèzes 1999; Driskell et al. 1994). Mental practice activates nearly the same neuronal circuitry as actual movements do, namely, regions in the premotor and the posterior parietal cortex, the basal ganglia, and the cerebellum (Jackson et al. 2001; Jeannerod 2001; Munzert et al. 2009). This approach has repeatedly been shown to improve motor performance; for instance, it optimizes arm kinematics (Pascual-Leone et al. 1996; Gentili et al. 2010) and increases muscle strength (Yue and Cole 1992; Ranganathan et al. 2004). These changes of performance are paralleled by changes of the underlying neuronal circuitry (Pascual-Leone et al. 1996; Lacourse et al. 2005).

Several key ingredients of successful mental practice have been proposed in literature, and the PETTLEP approach (Holmes and Collins 2001) combines probably the most promising ones. The name stands for *physical* (imagined body postures should be the same as in the actual task), *environment* (imagined architecture and furniture should be the same as in the actual task), *task* (movements, muscle forces, heart rate, etc., should be the same as in the actual task), *timing* (imagined movements should have the same timing as in actual task, no slow motion), *learning* (imagery should become more detailed as practice progresses), *emotion* (the same joy and arousal as with the actual task but no negative feelings), and *perspective* (first- or second-person perspective, as in the actual task). Thus in effect, PETTLEP aims to maximize the functional equivalence between imagined and actual tasks. To achieve this goal, PETTLEP can be combined with scripting (Immenroth et al. 2007). In this approach, experimenter and trainees observe, discuss, and then write down the sensory, motor, and affective aspects of the skill, and the outcome is subsequently broken down into key components identified by verbal labels. For training, the labels are called up and the pertinent imagery is instantiated in proper sequence.

The benefits of mental training can be understood within the framework of "internal models." It is though that our sensorimotor system uses forward internal models to mimic the causal flow of physical process and predict the future state of the body (e.g., positions, velocities) based on its current state, sensory feedback, and motor commands (Miall and Wolpert 1996). These models can be accessed and refined during mental practice even though the sensory feedback of a real movement is not available and even though refinement is probably less precise and accurate than it is in the presence of such feedback—despite its noisiness and delays (Wolpert et al. 1995).

According to the above considerations, it seems feasible to prepare future astronauts for physical exercise in weightlessness by mental training, thus replacing some of the present ground-based, underwater, and parabolic flight training. Participants would watch video footage of fellow astronauts who unstow, set up, use, and stow exercising equipment aboard the ISS and would mentally summon the movements, visual and proprioceptive sensations, as well as emotions during such activities, and an imagery script would then be set up with the astronaut's active participation, taking into account the seven components of the PETTLEP model. Important benefits of this training approach are that mental practice is fast and easy and that can be exercised anytime and anywhere, with no need for sparse and expensive facilities.

References

Aglioti S, DeSouza JFX, Goodale MA (1995) Size-contrast illusions deceive the eye but not the hand. Curr Biol 5:679–685

Baroni G, Ferrigno G, Rabuffetti M, Pedotti A, Massion J (1999) Long-term adaptation of postural control in microgravity. Exp Brain Res 128:410–416

Berger M, Mescheriakov S, Molokanova E, Lechner-Steinleitner S, Seguer N, Kozlovskaya I (1997) Pointing arm movements in short- and long-term spaceflights. Aviat Space Environ Med 68:781–787

Bloomberg J, Bock O (2012) Adaptation to weightlessness. In: Seel N (ed) Encyclopedia of the sciences of learning. Springer, Berlin, pp 102–103

Bock O (1998) Problems of sensorimotor coordination in weightlessness. Brain Res Rev 28:155–160

Bock O, Züll A (2013) Characteristics of grasping movements in a laboratory and in an everyday-like context. Hum Mov Sci 32:249–256

Bock O, Howard IP, Money KE, Arnold KE (1992) Accuracy of aimed arm movements in changed gravity. Aviat Space Environ Med 63:994–998

Bock O, Fowler B, Comfort D (2001) Human sensorimotor coordination during spaceflight. An analysis of pointing and tracking responses during the "Neurolab" space shuttle mission. Aviat Space Environ Med 72:877–883

Bock O, Fowler B, Jüngling S, Comfort D (2003) Visual-motor coordination during spaceflight. In: Buckey JC, Homick JL (eds) The Neurolab space mission: neuroscience res in space. NASA, Houston, TX, pp 83–89
Bock O, Weigelt C, Bloomberg JJ (2010) Cognitive demand of human sensorimotor performance during an extended space mission: a dual-task study. Aviat Space Environ Med 81:819–824
Bock O, Schott N, Papaxanthis C (2015) Motor imagery: lessons learned in movement science might be applicable for spaceflight. Front Syst Neurosci 9:75
Bridgeman B, Lewis S, Heit G, Nagle M (1989) Relation between cognitive and motor-oriented systems of visual position perception. J Exp Psychol 5:692–700
Burton AW, Miller DE (1998) Movement skill assessment. Human Kinetics, Champaign, IL
Campbell W, Buckey JC Jr, Kirkpatrick AW (2005) Animal surgery during spaceflight on the Neurolab shuttle mission. Aviat Space Environ Med 76:589–593
Cheron G, Leroy A, Palmero-Soler E, de Saedeleer C, Bengoetxea A, Cebolla A, Vidal M, Dan B, Berthoz A, McIntyre J (2014) Gravity influences top-down signals in visual processing. PLoS One 9:e82371
Clément G, Lestienne F (1988) Adaptive modifications of postural attitude in conditions of weightlessness. Exp Brain Res 72:381–389
Clément G, Ngo-Anh JT (2013) Space physiology II: adaptation of the central nervous system to space flight—past, current, and future studies. Eur J Appl Physiol 113:1655–1672
Clément G, Gurfinkel VS, Lestienne F, Lipshits MI, Popov KE (1984) Adaptation of postural control to weightlessness. Exp Brain Res 57:61–72
Clément G, Gurfinkel VS, Lestienne F, Lipshits MI, Popov KE (1985) Changes of posture during transient perturbations in microgravity. Aviat Space Environ Med 56:666–671
Clément G, Lathan C, Lockerd A (2008) Perception of depth in microgravity during parabolic flight. Acta Astronaut 63:828–832
Clément G, Skinner A, Richard G, Lathan C (2012) Geometric illusions in astronauts during long-duration spaceflight. Neuroreport 23:894–899
Dalecki M, Bock O, Schulze B (2012) Cognitive impairment during 5 m water immersion. J Appl Physiol 113:1075–1081
de Witt JK, Cromwell RL, Ploutz-Snyder LL (2014) Biomechanics of treadmill locomotion on the International Space Station. NASA Human Research Program Investigators Workshop, Galveston, TX
Decety J, Grèzes J (1999) Neural mechanisms subserving the perception of human actions. Trends Cogn Sci 3:172–178
Driskell J, Copper C, Moran A (1994) Does mental practice enhance performance? J Appl Psychol 79:481–492
Fitts PM (1954) The information capacity of the human motor system in controlling the amplitude of movement. J Exp Psychol 47:381–391
Fowler B, Comfort D, Bock O (2000) A Review of cognitive and perceptual-motor performance in space. Aviat Space Environ Med 71:A66–A68
Friederici AD, Levelt WJM (1990) Spatial reference in weightlessness. Perceptual factors and mental representations. Percept Psychophys 47:253–266
Gentili R, Han CE, Schweighofer N, Papaxanthis C (2010) Motor learning without doing: trial-by-trial improvement in motor performance during mental training. J Neurophysiol 104:774–783
Glasauer S, Mittelstaedt H (1992) Determinants of orientation in microgravity. IAA Man Space Symp 27:1–9
Glasauer S, Mittelstaedt H (1998) Perception of spatial orientation in microgravity. Brain Res Rev 28:185–193
Goodale MA, Meenan JP, Bulthoff HH, Nicolle DA, Murphy KJ, Racicot CI (1994) Separate neural pathway for the visual analysis of object shape in perception and prehension. Curr Biol 4:604–610
Harm DL, Parker DE (1993) Perceived self-orientation and self-motion in microgravity, after landing and during preflight adaptation training. J Vestib Res 3:297–305

Hermsdörfer J, Marquardt C, Philipp J, Zierdt A, Nowak D, Glasauer S, Mai N (2000) Moving weightless objects. Grip force control during microgravity. Exp Brain Res 132:52–64
Holmes PS, Collins DJ (2001) The PETTLEP approach to motor imagery: a functional equivalence model for sport psychologists. J Appl Sport Psychol 13:60–83
Human Integration Design Handbook (2010) NASA-SP-2010-3407. Washington, DC
Immenroth M, Bürger T, Brenner J, Nagelschmidt M, Eberspächer H, Troidl H (2007) Mental training in surgical education: a randomized controlled trial. Ann Surg 245:385
Jackson PL, Lafleur MF, Malouin F, Richards C, Doyon J (2001) Potential role of mental practice using motor imagery in neurologic rehabilitation. Arch Phys Med Rehabil 82:1133–1141
Jeannerod M (2001) Neural simulation of action: a unifying mechanism for motor cognition. Neuroimage 14:S103–S109
Kornilova LN (1997) Orientation illusions in spaceflight. J Vestib Res 7:429–439
Lackner JR, DiZio P (1993) Multisensory, cognitive, and motor influence on human spatial orientation in weightlessness. J Vestib Res 3:361–372
Lackner JR, DiZio P (1996) Motor function in microgravity: movement in weightlessness. Curr Opin Neurobiol 6:744–750
Lackner JR, Graybiel A (1979) Parabolic flight: loss of sense of orientation. Science 206:1105–1108
Lacourse MG, Orr ELR, Cramer SC, Cohen MJ (2005) Brain activation during execution and motor imagery of novel and skilled sequential hand movements. Neuroimage 27:505–519
Lipshits M, McIntyre J (1999) Gravity affects the preferred vertical and horizontal in visual perception of orientation. Neuroreport 10:1085–1089
Manzey D, Lorenz B, Schiewe A, Finell G, Thiele G (1995) Dual-task performance in space. Results from a single-case study during a short-term space mission. Hum Factors 37:667–681
McNay E, Willingham D (1998) Deficit in learning of a motor skill requiring strategy, but not of perceptual motor recalibration, with aging. Learn Mem 4:411–420
Miall RC, Wolpert D (1996) Forward models for physiological motor control. Neural Netw 9:1265–1279
Mishkin M, Ungerleider LG, Macko KA (1983) Object vision and spatial vision: two cortical pathways. Trends NeuroSci 6:414–417
Munzert J, Lorey B, Zentgraf K (2009) Cognitive motor processes: the role of motor imagery in the study of motor representations. Brain Res Rev 60:306–326
Navon D, Gopher D (1979) On the economy of the human-processing system. Psychol Rev 86:214–255
Nowak DA, Hermsdörfer J, Philipp J, Marquardt C, Glasauer S, Mai N (2001) Effects of changing gravity on anticipatory grip force control during point-to-point movements of a hand-held object. Motor Control 5:231–253
Oman C (2007) Spatial orientation and navigation in microgravity. In: Mast F, Jäncke L (eds) Spatial processing in navigation, imagery and perception. Springer, Berlin, pp 369–387
Oman CM, Lichtenberg BK, Money KE, McCoy RK (1986) M.I.T./Canadian vestibular experiments on the Spacelab-1 mission. 4. Space motion sickness: symptoms, stimuli, and predictability. Exp Brain Res 64:316–334
Panait L, Merrell RC, Rafiq A, Dudrick SJ, Broderick TJ (2006) Virtual reality laparoscopic skill assessment in microgravity. J Surg Res 136:198–203
Pascual-Leone A, Wassermann EM, Grafman J, Hallett M (1996) The role of the dorsolateral prefrontal cortex in implicit procedural learning. Exp Brain Res 107:479–485
Rafiq A, Broderick TJ, Williams DCR, Jones JA, Merrell RC (2005) Assessment of simulated surgical skills in parabolic microgravity. Aviat Space Environ Med 76:385–391
Ranganathan VK, Siemionow V, Liu JZ, Sahgal V, Yue GH (2004) From mental power to muscle power—gaining strength by using the mind. Neuropsychologia 42:944–956
Redding GM, Wallace B (1996) Adaptive spatial alignment and strategic perceptual-motor control. J Exp Psychol Hum Percept Perform 22:379–394

Robert G, Hockey J (1997) Compensatory control in the regulation of human performance under stress and high workload: a cognitive-energetical framework. Biol Psychol 45:73–93
Schmitt HH, Reid DJ (1985) Anecdotal information on space adaptation syndrome. NASA Johnson Space Center, Houston, TX
Schneider S, Brümmer V, Carnahan H, Kleinert J, Piacentini MF, Meeusen R, Strüder HK (2010) Exercise as a countermeasure to psycho-physiological deconditioning during long-term confinement. Behav Brain Res 211:208–214
Steinberg F, Bock O (2012) Influence of cognitive functions and behavioral context on grasping kinematics. Exp Brain Res 225:387–397
Steinberg F, Bock O (2013) Context dependence of manual grasping movements in near weightlessness. Aviat Space Environ Med 84:467–472
Tafforin C, Campan R (1994) Ethological experiments on human orientation behavior within a three-dimensional space-in microgravity. Adv Space Res 14:415–418
Tafforin C, Thon B, Guell A, Campan R (1989) Astronaut behavior in an orbital flight situation. Preliminary ethological observations. Aviat Space Environ Med 60:949–956
Thornton WE, Hoffler GW, Rummel JA (1977) Anthropometric changes and fluid shifts. In: Johnson RS, Dietlein LF (eds) Biomedical results from Skylab. NASA SP-377
Vidal M, Amorim M, Berthoz A (2004) Navigating in a virtual three-dimensional maze: how do egocentric and allocentric reference frames interact? Cogn Brain Res 19:244–258
Villard E, Garcia-Moreno FT, Peter N, Clément G (2005) Geometric visual illusions in microgravity during parabolic flight. Neuroreport 16:1395–1398
Watt DGD (1997) Pointing at memorized targets during prolonged microgravity. Aviat Space Environ Med 68:99–103
White RJ, Averner M (2001) Humans in space. Nature 409:1115–1118
Wolpert D, Ghahramani Z, Jordan M (1995) An internal model for sensorimotor integration. Science 269:1880–1882
Yue G, Cole KJ (1992) Strength increases from the motor program: comparison of training with maximal voluntary and imagined muscle contractions. J Neurophysiol 67:1114–1123

Chapter 2
Adaptation of Cartilage to Immobilization

A.-M. Liphardt, G.-P. Brüggemann, and A. Niehoff

Abstract Articular cartilage is essential for unconfined function of the musculoskeletal system. The effects of immobilization on hyaline cartilage have been investigated for many decades in cell and animal models, and it is known that normal mechanical loading, as experienced in daily life, is essential for cartilage health. Because of the slow rate of metabolism of cartilage, the time line for intervention experiments needs to be longer than for other skeletal tissues and the regenerative capacity of the cartilage is very limited, once degradation occurs. Thus, performing unloading experiments in healthy humans is difficult. A few studies have been performed in patient cohorts that experienced unloading due to injury, and the results suggest that human cartilage health is negatively affected by unloading.

Space flight research offers a unique opportunity to investigate musculoskeletal tissue adaptation to immobilization in either bed rest or Space flight experiments. Data on cartilage health are sparse but suggest that it is necessary to assess the risk of cartilage deconditioning during extensive human Space travel. Results from this context offer the unique possibility to broaden our understanding of the role of mechanical loading for tissue health.

Keywords Articular cartilage • Immobilization • Space flight • Bed rest • Mechanical loading • Chondrocytes • Osteoarthritis

---

A.-M. Liphardt (✉)
Institute of Biomechanics and Orthopaedics, German Sport University Cologne, Köln, Germany

Department of Internal Medicine 3 – Rheumatology and Immunology, Friedrich-Alexander-Universität Erlangen-Nürnberg (FAU), Erlangen, Germany
e-mail: Liphardt@dshs-koeln.de

G.-P. Brüggemann • A. Niehoff (✉)
Institute of Biomechanics and Orthopaedics, German Sport University Cologne, Köln, Germany

Cologne Center for Musculoskeletal Biomechanics, Medical Faculty, University of Cologne, Köln, Germany
e-mail: Brueggemann@dshs-koeln.de; Niehoff@dshs-koeln.de

S. Schneider (ed.), *Exercise in Space*, SpringerBriefs in Space Life Sciences,
DOI 10.1007/978-3-319-29571-8_2

2.1 Introduction

Healthy articular cartilage is the prerequisite for proper joint function and thus for unlimited physical activity. In synovial joints, articular cartilage serves a variety of functions including transferring and distributing forces and allowing joint movement between bone surfaces with minimal friction (Herzog 2007). Degradation of articular cartilage leads to osteoarthritis (OA), which is one of the leading causes of chronic disability worldwide (Scotece and Mobasheri 2015) (Fig. 2.1). OA is a complex disorder, and a variety of genetic, biomechanical, biochemical, and also physical factors are thought to play a role in its manifestation (Bedson et al. 2005). It cannot be cured, and the clinical characteristics of the end-stage disease are various degrees of joint pain, stiffness, dysfunction, and deformity as well as the radiographic manifestations of joint space narrowing, subchondral sclerosis, and osteophyte formation (Pearle et al. 2005). At present the ultimate treatment of severe OA is joint replacement. Understanding the basic science of cartilage and the role of mechanical loading or immobilization for homeostasis of the tissue is important to gain more insights into the mechanisms that trigger degenerative processes in cartilage. Mechanical loading is an ordinary characteristic of living in the gravity environment of the Earth. Even though our lifestyle is continuously changing toward more sedentary behavior, our musculoskeletal system is adapted to life in a gravity environment and needs mechanical stimuli for healthy development. The tissue characteristics and condition of articular cartilage at which mechanical loading no longer serves as a stimulus for cartilage maintenance but instead leads to increased degradation are still unknown. Also the role of immobilization for cartilage health is not well understood. Immobilization of healthy cartilage has been investigated in cell and animal studies and also in humans and will be discussed in this review. The latter is difficult, and most data available resulted from patient studies where immobilization was a result of an injury or the treatment of the symptoms. Space flight research offers the opportunity to investigate the effect of unloading on the human body in healthy individuals.

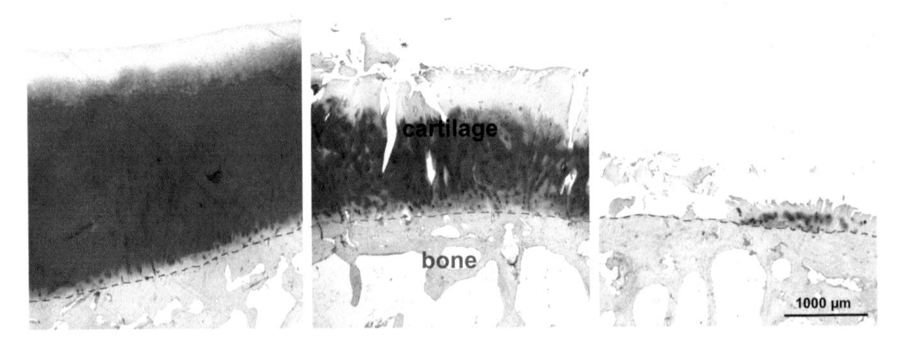

Fig. 2.1 Histology of healthy (*left*) and osteoarthritic (*middle* and *right*) articular cartilage of human femoral condyles. Sections were stained with safranin O for the detection of proteoglycans

Results obtained from such studies can be translated to clinical day-to-day situations, for instance, when patients are (partially) immobilized for bone fractures or other diseases.

2.2 Immobilization and Unloading in a Gravity Context

In the following, the terms *immobilization* and *unloading* are used in order to describe a condition/intervention for the investigated organism that is characterized by a reduction of mechanical loading compared to normal conditions in daily life where mechanical loading and/or joint motion occurs during walking, running, etc. The measures that are taken in different experiments to achieve the reduced mechanical loading can be various. They range from simple bed rest, where subjects spend their day bed ridden in a supine position, casting/splinting of a joint, or the use of crutches to unload one limb. Animal models, e.g., temporal paralysis of muscle using substances like botulinum toxin, have also been used in the past as well as hind limb unloading through tail fixation. Also, patient cohorts have been investigated where unloading of the lower limbs was used in the treatment of a disease or as a result of injury or disease. In general investigating the effects of immobilization/unloading is focused on parts of the body that are normally exposed to mechanical loading (e.g., lower limbs as compared to upper limbs or trunk) as the effect of the interventions is expected to be greatest in those locations. However, these types of models do not include the impact of gravitation, as on Earth it is more complex to elicit, especially when investigating animals or human subjects. Thus, if a patient is walking on crutches or an animal is subject to hind limb unloading, the mechanical loading through, e.g., muscle contraction of the respective joint/limb, is reduced, but gravity is still acting on the body, e.g., blood flow and fluid distribution are still acting under 1G conditions.

Gravity plays a major role in our daily life on Earth, and the evolution of the human body was strongly influenced by the gravitational force. When humans travel to Space, the body needs to adapt to an environment with neglectable gravitation and greatly reduced mechanical loading. The degree of tissue adaptation is greatly dependent on the location in the skeleton (LeBlanc et al. 2000) as especially lower limb joints experience significantly less mechanical loading as compared to daily life on Earth.

Immobilization using bed rest has been shown to be a highly feasible model to simulate effects of microgravity on the musculoskeletal system (Pavy-Le Traon et al. 2007). This has generally been accepted as the standard model for simulating the effects of microgravity for many physiological systems, particularly the musculoskeletal system, and has been used for research purposes since the 1960s. Kakurin et al. suggested that 6° head-down-tilt bed rest (6° HDT bed rest) simulates the effects of microgravity on the human body best (Pavy-Le Traon et al. 2007; Kakurin et al. 1976) as it also leads to a comparable fluid shift as in microgravity. Since then 6° HDT bed rest has been used as the gold standard to induce the effects of microgravity on the human organism. Water immersion provides an additional model to simulate the conditions of microgravity in Space flight studies as it can simulate the lower gravitational forces acting on the human body. However, a very complex experiment setting, which is not regularly available, is needed to apply this intervention, and thus, studies using water immersion are rare.

Cell experiments in the context of unloading research comprise a special situation, as because of their size it is easier to mimic microgravity using special experimental setups which will be described later.

In summary, studying the effects of immobilization on living organisms is difficult to do as being physical active and mechanical loading is generally a natural task. Different models are available to expose animals and also human subjects to defined lengths of immobilization. Traveling to Space includes the removal of gravitation, which is very difficult to mimic on Earth. Thus, Space flight offers a unique unloading model to research the impact unloading has on tissue health.

2.3 Structure and Function of Cartilage

Phylogenetically, cartilage is a very primitive tissue and represents the primary tissue of the skeleton (Carter et al. 2004). In the adult human skeleton, cartilage is present at the surface of the articulating bones but also in the walls of the thorax, larynx, trachea, bronchi, nose, and ears and the base of the skull (Moss and Moss-Salentijn 1983). The young small cartilage cells are called chondroblasts and are derived from mesenchymal stem cells. In mature cartilage, the cells are called chondrocytes and are larger in size. Chondrocytes are responsible for the continuous turnover of the components of the extracellular matrix (ECM).

Depending on the matrix composition and organization, mature cartilage can be distinguished into three types of cartilage which have very different tissue characteristics: yellow fibrocartilage, white fibrocartilage, and hyaline cartilage (Warwick and Williams 1973). The ECM of yellow fibrocartilage has an extensive network of elastin, which is a very compliant structural protein. This results in an easy deformability of yellow fibrocartilage structures like the external ears or epiglottis. White fibrocartilage is characterized by a very dense network of collagen fiber bundles with ovoid-shaped chondrocytes. The high fiber content provides this type of cartilage with tensile strength and toughness greater than that of hyaline cartilage, and it is found in locations where the tissue experiences high compressive stresses in one direction and tensile stresses in another direction, such as in the intervertebral disc and joint menisci (Carter et al. 2004). The effects of immobilization on fibrocartilage have been studied investigating the role of mechanical unloading on the intervertebral disc (Stokes and Iatridis 2004). In this review, the authors conclude that besides overuse, immobilization may play an important role in intervertebral disc degeneration as it may lead to altered material properties due to structure changes. More recent studies investigated the effects of bed rest on intervertebral disc morphology and describe changes due to bed rest that do not recover 5 months after bed rest (Belavy et al. 2012a, b). However, in the scope of this article, the focus is on the effects of immobilization on hyaline cartilage.

Instead of the large fiber bundles, hyaline cartilage has a more homogeneous structure of fine collagen fibrils and fibers. It appears between articulating joints, in the nose, and the xiphoid process of the sternum. With aging, hyaline cartilage is prone to ossification or calcification (Carter and Beaupre 2001). Articular cartilage has a very low rate of metabolism and can thus exist as a bradytrophic tissue without vascular canals. Additionally, it also has no nerves or lymphatic system. Cartilage nutrition thus happens primarily through diffusion from the surrounding synovial fluid within the joint and the synovial membrane. The low rate of metabolism also accounts for the very limited capacity of this highly specialized tissue to regenerate.

The composition and morphology of the articular cartilage is optimized for load-bearing function and characterized by the ability to sustain high and repetitive mechanical loads occurring during daily living. The unique mechanical properties of the cartilage are related to its multiphase composition and the structure of its highly organized ECM (Pearle et al. 2005). The main ECM components are collagen II and non-collagenous proteins, such as the proteoglycan aggrecan, and other glycoproteins. In total, cells only comprise 5 % of cartilage wet weight; nevertheless, chondrocyte metabolism is responsible for the maintenance of a stable and abundant ECM, and the balance between anabolism and catabolism of the matrix is crucial for articular cartilage homeostasis (Pearle et al. 2005).

Articular cartilage is organized in a four-layer system: superficial (or tangential) zone, middle (or transitional) zone, deep zone, and the zone of calcified cartilage. The different layers are characterized by specific cell size, collagen fibril orientation, and composition, which results in differences in the mechanical properties for each zone (Carter and Wong 2003). Briefly, the superficial zone is the articulating surface and in healthy cartilage provides a smooth gliding surface. It has the highest collagen content of the zones with collagen fibrils in a highly ordered alignment parallel to the articular surface (Hung and Mow 2012). Chondrocytes in this layer are characterized by an elongated appearance and preferentially express proteins that have lubricating and protective functions and secrete relatively little proteoglycan (Wong et al. 1996). The middle zone has a higher compressive modulus than the superficial zone, and the collagen fibrils in this zone are less organized, thicker fibers which are packed loosely and aligned obliquely to the surface. Chondrocytes in this layer are more rounded than in the superficial layer. In the deep zone, collagen fibrils are of large diameter and oriented perpendicular to the articular surface. This layer contains the highest proteoglycan and lowest water concentration and has the highest compressive modulus. The chondrocytes are typically arranged in columnar fashion parallel to the collagen fibers and perpendicular to the joint line. Finally, the tidemark separates the deep zone from the calcified cartilage, which rests directly on the subchondral bone. The calcified cartilage contains small cells in a chondroid matrix speckled with apatitic salts (Hung and Mow 2012).

The health and disease of the ECM is best understood when it is viewed as a biphasic structure (Pearle et al. 2005). Collagen and proteoglycans represent the solid phase, while the fluid phase is composed of water and ions. Proteins, lipids, phospholipids, and other collagens only make up a minor part of the tissue (Pearle et al. 2005). The solid component of the cartilage is composed primarily of a network of collagen fibrils maintained in a specific spatial arrangement by proteoglycan aggregates. Type II collagen contributes to the shear and tensile properties of the tissue. Like all collagens, type II collagen contains a characteristic triple helix structure. In the solid matrix, the collagen molecules line up in a staggered end-to-end and side-to-side fashion to form fibrils with holes and overlaps. Intra- and intermolecular cross-linking of the collagen fibrils serves to stabilize the matrix. In articular cartilage, the chief structural proteoglycan is aggrecan, which consists of a long protein core with up to 100 chondroitin sulfate and 50 keratan sulfate glycosaminoglycan chains (Heinegard 2009). These aggrecan molecules bind via a link protein on the protein core to a hyaluronate molecule, which form a backbone with palisading aggrecan molecules; this macromolecular complex is known as the proteoglycan aggregate and can reach the dimension of 30–75 million daltons (Pearle et al. 2005). Proteoglycans resist compression, and due to the high electronegativity of the aggrecan molecules, cartilage is a very hydrophilic tissue that swells in water. The interaction between the proteoglycan molecules and collagen fibrils creates a fiber-reinforced composite solid matrix. The proteoglycans are entangled and compacted within the collagen interfibrillar space, which helps to maintain a porous-permeable solid matrix and determines the movement of the fluid phase of the matrix (Heinegard 2009). The low permeability of the solid phase is due to high frictional resistance to fluid flow and causes a high interstitial fluid pressurization in the fluid phase. These two factors establish both the stiffness and the viscoelastic properties of cartilage (Felson et al. 2000). Collagen molecules in the fibrous system are thought to be under tension, while they resist the osmotic pressure which is increased when additional compressive load, through mechanical loading, is put on cartilage. The magnitude of tensile and shear strength of cartilage is thus attributed to collagen content and structure (Carter et al. 2004).

Water is the most abundant component of articular cartilage, accounting for 65–80 % of its wet weight (Mankin et al. 1999). The majority of water is contained within the interstitial intrafibrillar space created by the collagen–proteoglycan solid matrix and, as mentioned earlier, held in place by the negative charge on the proteoglycans. Water content is related to the *Donnan* osmotic pressure generated by the fixed negative charges on the proteoglycans (Maroudas 1976). The fluid phase provides the matrix with its viscoelastic properties, the time dependence, reversible deformability, and ability to dissipate load. Hydraulic pressure provides a significant component of load support of the cartilage, which protects and stress shields the solid phase of the matrix from much of the load.

2.4 Impact of Mechanical Loading on Cartilage Development, Growth, and Health

There is strong evidence from theoretical models (Carter and Wong 1988a; Wong et al. 1999; Wong and Carter 1988) and in vitro and in vivo studies that mechanobiological factors are essential for the development, health, and maintenance of articular cartilage. The role of mechanobiological factors for cartilage health can be investigated at different levels of consideration that range from the organ level over tissue and cellular level to the molecular level. Depending on the level of consideration, the parameters that can be investigated vary (Carter et al. 2004). Advances in imaging and measuring techniques continuously allow more detailed analysis at all levels of resolution, and with more detailed analytic methods, new questions arise. Thus, understanding the impact of mechanical loading on tissue development and aging tissue is a continuously evolving process.

Animal studies have shown that musculoskeletal movement is necessary for proper joint development, and in the absence of muscle activity, functional joint development is not possible (O'Rahilly and Gardner 1978; Kahn et al. 2009; Murray and Drachman 1969). Tissue stresses caused by joint movement through muscular activity affect histomorphogenesis starting in embryonic and fetal development resulting in a process of self-design of the skeleton (Nowlan et al. 2007; Kahn et al. 2009; Lelkes 1958; Shwartz et al. 2013; Pitsillides 2006). Mechanical loading during developmental processes modulates cartilage growth rates and thus influences skeletal morphogenesis resulting in shaping joints with improved biomechanical characteristics (Carter et al. 2004). There is no formation of a joint space between the tibia and femur when the limbs of an embryo are not moved (Lelkes 1958). Also, there is evidence that hydrostatic tissue pressure caused by loading is chondrogenic, maintains the cartilage phenotype, and locally reduces cartilage growth (Carter et al. 2004). Furthermore, there are various musculoskeletal diseases which are influenced through mechanical factors in the prenatal phase. For example, the risk to develop hip dysplasia is highly increased when the hip abduction of the fetus is limited due to an unfavorable position in the womb (Shefelbine and Carter 2004).

In addition, mechanobiological factors are thought to cause changes in articular cartilage thickness distributions in a joint throughout life (Carter and Beaupre 2001), and cartilage, like other biological tissues including skeletal muscle and bone, is sensitive to loading and disuse (Smith et al. 1992; Johnson 1998). Repetitive high-impact loading has been implicated to enhance the risk of cartilage damage resulting in osteoarthritis (OA) (Buckwalter 2003; Buckwalter and Martin 2004). Although the genetic background is an important factor in the development of OA (Newman and Wallis 2002), it is not clear how mechanical loading contributes to cartilage degeneration. From a clinical and health economical perspective, it is important to identify which types of mechanical loading and loading characteristic may contribute to what extent to cartilage health or development of OA. High-impact loading creates cracks in the superficial zone with subchondral changes

(Thompson et al. 1991), matrix damage followed by cell death (Jeffrey et al. 1995), and chondrocyte synthesis (Farquhar et al. 1996).

During joint articulation, forces and also contact areas in the joint vary greatly and can reach up to ten times body weight (Andriacchi et al. 1997). Under physiological conditions, cartilage is a highly stressed tissue and thus is characterized by specific load-carrying mechanisms. Healthy articular cartilage tends to be thickest in joints that experience high forces such as the knee. In addition, side differences in the muscle's cross-sectional area positively correlated with side differences in articular cartilage morphology (Eckstein et al. 2002b). Previous studies (Andriacchi et al. 2004) showed that for healthy knees, the ratio of medial to lateral cartilage is greater in individuals that have a larger peak knee adduction moment during walking, suggesting that cartilage is thicker in areas where load is greater. These results are supported by those from animal studies (Hamann et al. 2012, 2013; Kiviranta et al. 1987, 1988, 1992) in which articular cartilage increased in thickness by up to 19–23 % (Kiviranta et al. 1988) when high mechanical loads were applied. An increase in cartilage thickness with exposure to higher loads may be offset by a greater cartilage surface that may be caused by high physical activity during growth (Eckstein et al. 2002a, 2009). Changes in tibiofemoral cartilage thickness are not dose dependent suggesting that adult human cartilage properties may not be sensitive to or improve with training (Eckstein et al. 2005b). However, mechanical loading as an intervention (e.g., exercise) has still been suggested as an effective non-pharmacological treatment for restoring physical function and reducing OA symptoms (Griffin and Guilak 2005). Previous studies have demonstrated that chondrocytes can detect and response to mechanical loading by altering their metabolism (Davisson et al. 2002; Niehoff et al. 2008; Steinmeyer et al. 1997; Wong et al. 1999). This relationship has been shown to depend on both the character of the mechanical loading (Bachrach et al. 1995; Sah et al. 1989; Sauerland et al. 2003) and the localization of chondrocytes within the cartilage (Parkkinen et al. 1992; Wong et al. 1997). A better understanding of the relation between mechanical loading and cartilage response will contribute to the development of new strategies for both the prevention and treatment of cartilage degeneration.

2.5 Assessment of Cartilage Health In Vivo in Humans in Relation to Mechanical Loading

To analyze the effect of immobilization on articular cartilage in humans, suitable noninvasive methods are necessary that can measure tissue adaptation in an accurate and reproducible way, which is particularly important for prospective clinical trials. The least invasive methods are imaging techniques that quantitatively, and recently also qualitatively, enable the investigator to image cartilage tissue in vivo. Furthermore, imaging techniques have the advantage of being site specific. Hunter et al. provide a recent comprehensive overview of the latest imaging techniques in a review initiated by the Osteoarthritis Research Society International (OARSI) (Hunter et al. 2015).

Briefly, potential imaging techniques to illustrate articular cartilage comprise of X-ray, ultrasound, PET and SPECT/scintigraphy, computed tomography (CT), and magnet resonance imaging (MRI). X-ray indirectly offers the diagnosis of OA using joint space narrowing; however, this technique does not allow qualitative and quantitative analysis of cartilage health and is only suitable when cartilage degradation is already advanced. CT arthrography in combination with contrast agents allows the investigation of both quantitative and qualitative tissue properties (Bansal et al. 2010); however, radiation exposure and the invasive nature of injecting a contrast agent make it difficult to apply.

MRI is suggested to be the most appropriate imaging technique to assess cartilage health (Peterfy et al. 2008) that is becoming more available and also requires no exposure to radiation (Eckstein et al. 2006). The technological developments in high-resolution magnetic resonance imaging techniques have resulted in precise, accurate, and reproducible protocols to determine cartilage morphology in vivo (Eckstein 2004; Eckstein et al. 2005a) (Fig. 2.2). Also, qualitative investigation is possible when using T2 (Mosher and Dardzinski 2004) or T1rho mapping (Burstein et al. 2009) of cartilage or gadolinium-enhanced MRI of cartilage (dGEMRIC) (Bashir et al. 1997; Burstein et al. 2001). However, MRI imaging techniques have only advanced and developed in the last decade, and thus, data on qualitative imaging in the context of immobilization in humans is still sparse.

Besides these imaging techniques, several biochemical markers have been identified to monitor cartilage metabolism using biological samples. Biomarkers in general are defined as "a characteristic that is objectively measured and evaluated as an indicator of normal biological processes, pathogenic processes, or pharmacological responses to a therapeutic intervention" (Biomarkers and surrogate endpoints: preferred definitions and conceptual framework 2001). In the last few years, a variety of OA-related biochemical markers have been identified to monitor

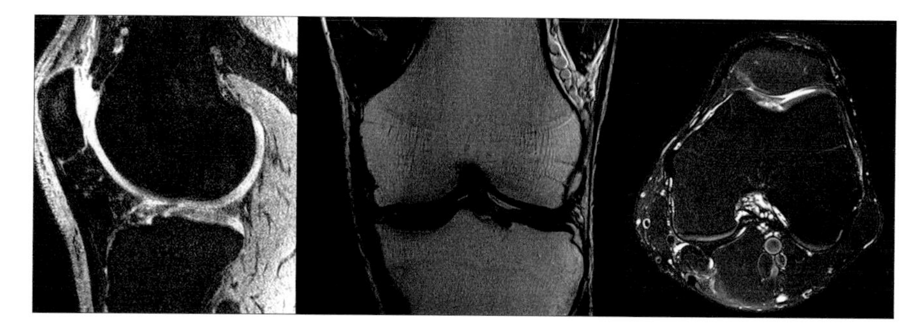

Fig. 2.2 Sagittal (*left*), coronal (*middle*), and transversal (*right*) MRI images of a human knee. Images can be used to determine cartilage morphology in vivo in a precise, accurate, and reproducible way

cartilage metabolism. Potential biological markers include matrix components and/or their breakdown products, cytokines, and proteases, and they can be quantified in serum, urine, or synovial samples (Bauer et al. 2006). These OA biomarkers can detect and predict disease progression and outcomes, but currently only a few biomarkers are validated (Kraus et al. 2010). The biomarkers of cartilage health have mainly been established in clinical environments and patients with diseases that are known to affect cartilage health, like OA but also rheumatoid arthritis (RA). Distinctly different patterns of these biomarkers have been observed in patient cohorts compared to healthy individuals. Biomarker levels can identify imbalanced catabolism and anabolism. However, measuring a biomarker response in urine and blood samples always reflects the systemic effect of a disease or intervention. This also means that an effect has to be large enough to show systemically and also that a biomarker response can only be as specific as the measured metabolite. In the following, a few examples of studies investigating a cartilage biomarker adaptation in the context of disease or exercise interventions will be given.

Cartilage oligomeric matrix protein (COMP) is a well-studied biomarker for cartilage degeneration (Neidhart et al. 1997). This pentameric non-collagenous protein found in the cartilage ECM is mainly a bridging molecule which influences the synthesis and remodeling of the ECM under physiological and pathophysiological conditions (Acharya et al. 2014). For example, it interacts with cartilage collagens and seems to play a role in fibril assembly and maintaining the mature collagen network (Hedbom et al. 1992; Rosenberg et al. 1998; Halasz et al. 2007; Johnson et al. 2004; Mann et al. 2004). Changes in serum COMP concentration may be an indicator of cartilage catabolism (Jordan 2005). Elevated COMP levels have been found for OA (Petersson et al. 1998; Saxne and Heinegard 1992), RA (Lindqvist et al. 2005; Mansson et al. 1995), and injured knees (Dahlberg et al. 1994). Furthermore, there is evidence that COMP is sensitive to mechanical loading. Mündermann et al. could describe biomarker kinetics of COMP before and after a 30-min walking exercise (Mundermann et al. 2005). Serum concentration of COMP increased immediately after the walking exercise and again 5.5 h after the stimulus. In a later study, the recovery characteristics of the biomarker response to a mechanical stimulus could be related to the risk of cartilage thinning (Erhart-Hledik et al. 2012). Increases in serum COMP have also been observed after running 30 min (Niehoff et al. 2010, 2011) or 10 km (Kim et al. 2009) as well as after a marathon (Neidhart et al. 2000; Kim et al. 2009) and an ultramarathon (Kim et al. 2009). The magnitude of the increase of the COMP concentration and the time until values recover seem to be related to running distance. The longer the distance, the higher the increase and the longer a recovery to baseline needs (Kim et al. 2009). A distinct increase in serum COMP was measured in the second half of the ultramarathon (Kim et al. 2007), and values stayed significantly higher for 4 days after the race with recovery to baseline only reached on day 6 after racing (Kim et al. 2009). The authors argue that cartilage destruction is increasing with increasing running distance. For serum COMP concentration, also the type of exercise seems to play an important role. While a 30-min running exercise leads to a significant increase in serum COMP concentration, 120 squads did not have an effect on serum COMP (Niehoff et al. 2010). Also 100 drop jumps lead to a similar increase in serum COMP concentration as 30-min running (Niehoff et al. 2011). In patients with OA, a weight-bearing exercise further increased the serum COMP concentration (Mundermann et al. 2009; Andersson et al. 2006). These findings support suggestions that cyclic mechanical loading may play an important role in regulating metabolic processes associated with the development (Wong et al. 1999; Carter and Wong 1988b), remodeling, and disease of biological tissues including cartilage. In particular, the turnover of COMP fragments may be an important mechanism for the regulation of tissue synthesis and degradation (Wong et al. 1999; Giannoni et al. 2003). The underlying mechanisms to explain COMP response to exercise interventions are not well understood, and more studies, especially prospective cohorts, are needed. Also, it is not known if and how these acute and short-term effects impact long-term adaptations of articular cartilage.

For the understanding of the impact and importance of mechanical loading for tissue health, observing the response of cartilage to unloading is essential. With this the full spectrum of mechanical loading, from zero to high mechanical loads, the respective tissue adaptations would be covered. In the following paragraphs, results of unloading/immobilization studies on different levels of tissue and organism size using various experimental setups are summarized.

2.6 Adaptations of Cartilage to Immobilization and Microgravity

2.6.1 *Cell Studies*

Observing the effects of unloading in cell metabolism is important in order to understand the biology of three-dimensional (3D) structures and the function of a tissue and the relevance of mechanical loading for both (Grimm et al. 2014). In a recent review, Grimm et al. summarized the advances of tissue engineering in the context of microgravity research (Grimm et al. 2014). It has been known for several decades that cells generally are sensitive to microgravity (Cogoli et al. 1984). Simulated microgravity can be applied to improve the 3D cell aggregation in vitro especially to investigate the early cellular events that initiate the formation of cartilage (Grimm et al. 2014). Furthermore, microgravity triggers many cells including chondrocytes to grow within 3D aggregates (Freed et al. 1997; Hsu et al. 2005). Because of the costly and timely nature of Space flight experiments, ground-based models have been developed in order to simulate microgravity. For cell culture experiments, microgravity can be simulated using several devices. In a fast rotating clinostat, cell suspensions rotate in tubes at a speed of 60–90 rpm (Briegleb 1992). The random positioning machine is a 3D clinostat where two frames rotate at randomized speeds and directions (Hoson et al. 1997). Chondrocytes are resistant to adverse effects and stress induced by microgravity (Freed et al. 1997; Klement and Spooner 1999; Montufar-Solis and Duke 1999).

Chondrocytes exposed to simulated microgravity using a random positioning machine change the ECM production and rearrange their cytoskeletal proteins (Grimm et al. 2014; Aleshcheva et al. 2013). Collagen type II, chondroitin sulfate, and aggrecan contents were higher in chondrocytes subjected to simulated microgravity as compared to control cells (Grimm et al. 2014).

Simulated microgravity bears the opportunity to culture chondrocytes with the aim to grow cartilage tissue without using scaffolds (Ulbrich et al. 2010).

2.6.2 Animal Models

Immobilization models on animals have been conducted for several decades and provide most of the knowledge we have to date regarding the adaptation of articular cartilage to immobilization. A variety of immobilization models have been applied, such as the tail suspension model (Tomiya et al. 2009; Leong et al. 2010), splinting (Jurvelin et al. 1986; LeRoux et al. 2001), external fixation (Hagiwara et al. 2009), postoperative immobilization (Roth et al. 1988), spinal cord injury (Moriyama et al. 2008), and muscle weakness through botulinum toxin (Rehan Youssef et al. 2009). In most of these studies, the animals are immobilized for a given time period, and parameters are compared to those of control animals that were free to move in their cages. Changes in cartilage due to immobilization were analyzed on the macroscopic level, and/or morphological, biochemical, or mechanical properties were examined.

An early study from Evans et al. (1960) demonstrated in a rat model that immobilization leads to cartilage matrix fibrillation, ulceration, and erosion. Rehan Youssef et al. (2009) demonstrated in a study using botulinum toxin type A to indirectly cause joint unloading through muscle weakness that cartilage degenerates in the absence of normal muscle function and movement. However, in other studies, immobilization did not cause macroscopic changes of the cartilage surface (Jurvelin et al. 1986; Moriyama et al. 2008). In addition, there are inconsistent results regarding the effect of unloading on cartilage thickness. Previous studies have demonstrated that immobilization decreased, increased, or had no affect on cartilage thickness (Haapala et al. 1999; Hagiwara et al. 2009; Jurvelin et al. 1986; LeRoux et al. 2001). The response of cartilage thickness to unloading is known to be dependent on the tissue site, e.g., patella, femur, or tibia; tissue region, e.g., anterior versus posterior; or cartilage depth indicating a different adaptation of the cartilage to the loading characteristics of the different immobilization models (LeRoux et al. 2001). The published results on the effects of mechanical unloading on changes in proteoglycan and collagen content in animals are contrasting. A study on young beagle dogs found decreases in proteoglycan concentration and compressive stiffness after 11 weeks of immobilization using a splint that immobilized the right hind limbs of the dogs at 90° flexion of the knee and ankle joint (Jurvelin et al. 1986). The instant elastic modulus of the cartilage at the femur, tibia, and patella was decreased by 17 %. This tissue softening was further confirmed in different studies (Haapala et al. 2000; Hagiwara et al. 2009; LeRoux et al. 2001). Further investigations of the same group showed that the immobilization-induced cartilage softening was not fully restorable (Haapala et al. 1999). In contrast, no changes in proteoglycan and collagen content, cartilage stiffness, or cartilage thickness after 4 and 8 weeks of immobilization were observed in a study by Setton et al. (1997). In the study they immobilized canine knees with a sling fixing the knee joint at 90° flexion and thus causing mechanical unloading but still allowing minimum motion in the joint (Setton et al. 1997). The discrepancy of these results compared to earlier studies may have been due to the fact that the type of immobilization played a role in the potential of the cartilage to recover (Behrens et al. 1989).

In a comprehensive study, Leong et al. (2010) demonstrated an increase of MMP-3 mRNA expression 6 h after immobilization using hind limb unloading in rats that lasted until 21 days of immobilization, suggesting that immobilization may trigger a catabolic gene expression response early after immobilization. Additionally, increased MMP-3 levels consequently contributed to increases in ADAMTS-5 expression and reduced safranin O staining. Further, the conducted spatial analysis of changes in MMP-3 gene and protein levels during immobilization revealed that upregulation not only varied among zones within a region of articular cartilage but also differed between condyles. In line with the earlier studies, Leong et al. (2010) suggest that reduced load and also overloading may share a common progression of cartilage degradation (Tetlow et al. 2001; Tetlow and Woolley 2001; Okada et al. 1992).

2.6.3 *Human Studies*

While applying unloading in cells or animal models is feasible, the difficulty increases with increasing size of the organism. In humans investigating the effect of unloading on articular cartilage is difficult, and even more so in healthy individuals. In this context Space flight-related models provide a unique possibility to investigate this particular response.

Due to the lack of noninvasive ethical immobilization methods, current knowledge of immobilization effects on articular cartilage is based on a few studies which have investigated the influence of mechanical unloading on articular cartilage in patient cohorts (Owman et al. 2014; Vanwanseele et al. 2002, 2003; Muhlbauer et al. 2000; Hinterwimmer et al. 2004; Hudelmaier et al. 2006). Here, only unloading after spinal cord injury, ankle fractures, or knee surgeries provides an opportunity to analyze the effects of no or reduced cartilage loading on tissue integrity. In paraplegic patients after spinal cord injury, cartilage thinning of up to 25 % after 24 months has been observed (Vanwanseele et al. 2003). In contrast to the knee, articular cartilage thickness of the humeral head was not affected through spinal cord injury (Vanwanseele et al. 2003). Hinterwimmer et al. (2004) investigated the effect of 7 weeks of partial load bearing after an ankle fracture on articular cartilage morphology in different knee compartments. The reported changes in articular cartilage thickness for the different compartments ranged from -2.9 ± 3.2 % for the patella to -6.6 ± 4.9 % for the medial tibia. Hudelmaier et al. (2006) detected a reduction in patellar cartilage thickness of 14 % but no changes at the tibia in a patient affected by 6 weeks of immobilization after a knee joint surgery. It is not known yet why the medial compartment seems more mechanically sensitive to reduced loading, but it has been suggested that differences in contact forces or pressures (Andriacchi et al. 2004; Bullough 2004) and differential contact with the menisci between the lateral and medial femoral condyles (Freeman and Pinskerova 2005) may be a causal factor for condyle-dependent differences in cartilage degradation (Leong et al. 2010). Since the load distribution in the joint is primarily transmitted through the medial compartment during weight-bearing activities (Thambyah 2007), in reduced loading conditions, this compartment may see a greater percent decrease in stress, resulting in the greater rate of degradation (Liphardt et al. 2009). A recent study of Owman et al. (2014) also used an ankle fracture model to analyze the long-term effect of joint unloading on the properties of the knee cartilage ECM by delayed gadolinium-enhanced MRI. The unloading of the knee joint for 6 weeks resulted in a broader T1Gd range indicating an adaptation of the matrix. In summary imaging to date suggests that both cartilage thickness and also cartilage ECM integrity in patients are sensitive to unloading.

Interestingly, changes in serum biomarkers of cartilage metabolism have been demonstrated after anterior cruciate ligament injury compared to uninjured matched controls indicating an alteration in cartilage turnover (Svoboda et al. 2013). However, studies on the response of biochemical markers to mechanical interventions are still sparse, and more studies are needed to better understand the impact of mechanical loading on biomarker concentration in vivo in humans and also on the meaning of alterations in biomarker concentration for the cartilage metabolism.

2.6.4 Space Flight-Related Research

The question whether joints are still functional after several months in microgravity is also essential for astronauts' health during Space travel and for rehabilitation regiments after Space flight. Currently astronauts are in Space for several months, mostly 5–6 months, and return to Earth where they are supported from a team of medical professionals in their readaptation to Earth conditions. A planned flight to Mars with an assumed duration of 2–3 years emphasizes the urgent need for research in the field of cartilage health under mechanical unloading as present during Space flight or immobilization, especially with regard to the abovementioned findings from spinal cord injured patients, where already 12 months after injury profound effects on cartilage morphology were found (Vanwanseele et al. 2002). A flight to Mars would exceed this presumably critical time window by far, and at the moment, the risk for joint health and joint function after a journey in microgravity for several years cannot be estimated. However, proper joint function is essential for locomotion in a gravity environment, and crew support would not be possible if crews land at other destinations than Earth. Thus, it is very important to assess the risk of microgravity for joint health and, more specifically, the ability to functionally load the joint again after many months in microgravity.

Although the effects of microgravity on bone and muscle have been studied extensively, given lack of appropriate noninvasive imaging technology until recently, little is known about the effects of immobilization on articular cartilage morphology and composition in humans in response to a longer stay in microgravity. Also the length of Space travel increased in the past decade with the establishment of the International Space Station (ISS). Before that, Space travel mostly happened for several days or weeks as compared to months.

Cartilage health of the lower limb joints has been investigated in microgravity analogue bed rest studies. Fourteen days of bed rest affected cartilage thickness at the knee and also the serum oligomeric matrix protein (COMP) concentrations (Liphardt et al. 2009). Furthermore, we have shown that COMP, matrix metalloprotease-3 (MMP-3), and matrix metalloprotease-9 (MMP-9) were sensitive to 5 and 21 days of bed rest indicating a critical time window for cartilage response to unloading after 2–3 weeks of unloading (Liphardt et al. 2015). Matrix metalloproteinases (MMPs) are a multi-member family of proteases with a wide range of substrates including extracellular components, cytokines, receptors, and cell motility factors (Yong et al. 2007; Morrison et al. 2009). They are involved in the degradation of the components of the cartilage matrix. For example, native collagen II is degraded by matrix metalloproteinases (MMP)-1, MMP-8, MMP-13, and MMP-14, and partially degraded collagen II is then further degraded by gelatinases, namely, MMP-2 and MMP-9 and stromelysin (MMP-3) (De Ceuninck et al. 2011). MMPs represent the main proteolytic enzyme group involved in remodeling the ECM and modifying cell–cell and cell–matrix interactions. Previous studies suggest that MMPs play a key role in regulating the balance of structural proteins of the articular cartilage matrix according to local mechanical demands (Monfort et al. 2006).

Applied countermeasures such as vibration training with (Liphardt et al. 2015) or without (Liphardt et al. 2009) additional resistive exercise could not compensate effects on cartilage metabolism. But future studies with prospective data collection are needed, to better understand the impact of immobilization on cartilage health. Since the risk of cartilage degeneration as a result of prolonged immobilization and microgravity is not known, the potential of exercise to counteract the initiation of cartilage degeneration in a healthy joint can only be estimated at the moment.

2.7 Summary and Conclusions

Articular cartilage is essential for unconfined movement of the musculoskeletal system. Because of its bradytrophic nature, diffusion is necessary for nutrition, and it has been shown that the absence of mechanical loading or muscle activity during developmental stages leads to malformation of cartilage in articular joints. The effects of immobilization on hyaline cartilage health have been investigated for many decades in cell experiments and animal models, and from these studies, it is well known that normal mechanical loading, as the joints are exposed to in daily life, is essential for cartilage health. Immobilizations can lead to cartilage softening and changes in the ECM that are not recovering when applying exercise interventions. Because of the slow rate of metabolism of cartilage, the time line for intervention experiments needs to be longer than for other skeletal tissues like the bone or muscle. Thus, performing similar unloading experiments in humans is difficult to do, even more so for healthy individuals without comorbidities. Results from the few studies that have been performed in patient cohorts that experienced long unloading periods of several weeks or months due to injury or illness suggest that unloading also negatively affects cartilage health in humans as they have shown that cartilage thickness decreases. In the future, development of imaging techniques, such as MRI, and validation of biomarkers of cartilage metabolism will allow for more accurate and also quantitative assessment of cartilage adaptations to immobilization in vivo in humans. Space flight research at different tissue levels offers unique technologies and experimental setups that allow the investigation of the adaptation of the musculoskeletal tissue to immobilization. Interestingly, in tissue engineering simulated microgravity may offer a possibility to cultivate cartilage aggregates for the treatment of cartilage degeneration. In this particular environment, cell growth in a 3D manner is possible. In human studies, integrative research approaches using bed rest and in Space flight experiments allow for comprehensive investigation of tissue adaptation on the morphological and also biological level. Data from these studies are still very sparse but suggest that cartilage is sensitive to unloading in healthy individuals and that it is necessary to assess the risk of cartilage deconditioning during extensive human Space travel. Data acquired in the Space flight context also offers a unique possibility to broaden our understanding of the role of mechanical loading for tissue health. This is especially interesting for conditions where tissue injury or disease subsequently also leads to immobilization because of pain or weakness which then results in a very sedentary lifestyle.

References

Acharya C, Yik JH, Kishore A, Van Dinh V, Di Cesare PE, Haudenschild DR (2014) Cartilage oligomeric matrix protein and its binding partners in the cartilage extracellular matrix: interaction, regulation and role in chondrogenesis. Matrix Biol 37:102–111. doi:10.1016/j.matbio.2014.06.001

Aleshcheva G, Sahana J, Ma X, Hauslage J, Hemmersbach R, Egli M, Infanger M, Bauer J, Grimm D (2013) Changes in morphology, gene expression and protein content in chondrocytes cultured on a random positioning machine. PLoS One 8(11):e79057. doi:10.1371/journal.pone.0079057

Andersson ML, Thorstensson CA, Roos EM, Petersson IF, Heinegard D, Saxne T (2006) Serum levels of cartilage oligomeric matrix protein (COMP) increase temporarily after physical exercise in patients with knee osteoarthritis. BMC Musculoskelet Disord 7:98. doi:10.1186/1471-2474-7-98

Andriacchi TP, Natarajan RN, Hurwitz DE (1997) Musculoskeletal dynamics, locomotion, and clinical applications. In: Mow VCH, Hayes WC (eds) Basic orthopaedic biomechanics, 2nd edn. Lippincott-Raven, Philadelphia, pp 31–68

Andriacchi TP, Mundermann A, Smith RL, Alexander EJ, Dyrby CO, Koo S (2004) A framework for the in vivo pathomechanics of osteoarthritis at the knee. Ann Biomed Eng 32(3):447–457

Bachrach NM, Valhmu WB, Stazzone E, Ratcliffe A, Lai WM, Mow C (1995) Changes in proteoglycan synthesis of chondrocytes in articular cartilage are associated with the time-dependent changes in their mechanical environment. J Biomech 28(12):1561–1569. doi:10.1016/0021-9290(95)00103-4

Bansal PN, Joshi NS, Entezari V, Grinstaff MW, Snyder BD (2010) Contrast enhanced computed tomography can predict the glycosaminoglycan content and biomechanical properties of articular cartilage. Osteoarthritis Cartilage 18(2):184–191. doi:10.1016/j.joca.2009.09.003

Bashir A, Gray ML, Boutin RD, Burstein D (1997) Glycosaminoglycan in articular cartilage: in vivo assessment with delayed Gd(DTPA)(2-)-enhanced MR imaging. Radiology 205 (2):551–558. doi:10.1148/radiology.205.2.9356644

Bauer DC, Hunter DJ, Abramson SB, Attur M, Corr M, Felson D, Heinegard D, Jordan JM, Kepler TB, Lane NE, Saxne T, Tyree B, Kraus VB (2006) Classification of osteoarthritis biomarkers: a proposed approach. Osteoarthritis Cartilage 14(8):723–727. doi:10.1016/j.joca.2006.04.001

Bedson J, Jordan K, Croft P (2005) The prevalence and history of knee osteoarthritis in general practice: a case-control study. Fam Pract 22(1):103–108. doi:10.1093/fampra/cmh700

Behrens F, Kraft EL, Oegema TR Jr (1989) Biochemical changes in articular cartilage after joint immobilization by casting or external fixation. J Orthop Res 7(3):335–343

Belavy DL, Armbrecht G, Felsenberg D (2012a) Evaluation of lumbar disc and spine morphology: long-term repeatability and comparison of methods. Physiol Meas 33(8):1313–1321. doi:10.1088/0967-3334/33/8/1313

Belavy DL, Armbrecht G, Felsenberg D (2012b) Incomplete recovery of lumbar intervertebral discs 2 years after 60-day bed rest. Spine (Phila Pa 1976) 37(14):1245–1251. doi:10.1097/BRS.0b013e3182354d84

Biomarkers Definitions Working Group (2001) Biomarkers and surrogate endpoints, preferred definitions and conceptual framework. Clin Pharmacol Ther 69(3):89–95. doi:10.1067/mcp.2001.113989

Briegleb W (1992) Some qualitative and quantitative aspects of the fast-rotating clinostat as a research tool. ASGSB Bull 5(2):23–30

Buckwalter JA (2003) Sports, joint injury, and posttraumatic osteoarthritis. J Orthop Sports Phys Ther 33(10):578–588. doi:10.2519/jospt.2003.33.10.578

Buckwalter JA, Martin JA (2004) Sports and osteoarthritis. Curr Opin Rheumatol 16(5):634–639

Bullough PG (2004) The role of joint architecture in the etiology of arthritis. Osteoarthritis Cartilage 12(Suppl A):S2–9

Burstein D, Velyvis J, Scott KT, Stock KW, Kim YJ, Jaramillo D, Boutin RD, Gray ML (2001) Protocol issues for delayed Gd(DTPA)(2-)-enhanced MRI (dGEMRIC) for clinical evaluation of articular cartilage. Magn Reson Med 45(1):36–41

Burstein D, Gray M, Mosher T, Dardzinski B (2009) Measures of molecular composition and structure in osteoarthritis. Radiol Clin North Am 47(4):675–686. doi:10.1016/j.rcl.2009.04.003

Carter DR, Beaupre GS (2001) Skeletal function and form – mechanobiology of skeletal development, aging, and regeneration, vol 1. Cambridge University Press, Cambridge

Carter DR, Wong M (1988a) Mechanical stresses and endochondral ossification in the chondroepiphysis. J Orthop Res 6(1):148–154

Carter DR, Wong M (1988b) The role of mechanical loading histories in the development of diarthrodial joints. J Orthop Res 6(6):804–816

Carter DR, Wong M (2003) Modelling cartilage mechanobiology. Philos Trans R Soc Lond B Biol Sci 358(1437):1461–1471

Carter DR, Beaupre GS, Wong M, Smith RL, Andriacchi TP, Schurman DJ (2004) The mechanobiology of articular cartilage development and degeneration. Clin Orthop Relat Res 427:S69–S77

Cogoli A, Tschopp A, Fuchs-Bislin P (1984) Cell sensitivity to gravity. Science 225 (4658):228–230

Dahlberg L, Roos H, Saxne T, Heinegard D, Lark MW, Hoerrner LA, Lohmander LS (1994) Cartilage metabolism in the injured and uninjured knee of the same patient. Ann Rheum Dis 53 (12):823–827

Davisson T, Kunig S, Chen A, Sah R, Ratcliffe A (2002) Static and dynamic compression modulate matrix metabolism in tissue engineered cartilage. J Orthop Res 20(4):842–848. doi:10.1016/S0736-0266(01)00160-7

De Ceuninck F, Sabatini M, Pastoureau P (2011) Recent progress toward biomarker identification in osteoarthritis. Drug Discov Today 16(9–10):443–449. doi:10.1016/j.drudis.2011.01.004

Eckstein F (2004) Noninvasive study of human cartilage structure by MRI. Methods Mol Med 101:191–217. doi:10.1385/1-59259-821-8:191

Eckstein F, Faber S, Muhlbauer R, Hohe J, Englmeier KH, Reiser M, Putz R (2002a) Functional adaptation of human joints to mechanical stimuli. Osteoarthritis Cartilage 10(1):44–50

Eckstein F, Muller S, Faber SC, Englmeier KH, Reiser M, Putz R (2002b) Side differences of knee joint cartilage volume, thickness, and surface area, and correlation with lower limb dominance—an MRI-based study. Osteoarthritis Cartilage 10(12):914–921

Eckstein F, Charles HC, Buck RJ, Kraus VB, Remmers AE, Hudelmaier M, Wirth W, Evelhoch JL (2005a) Accuracy and precision of quantitative assessment of cartilage morphology by magnetic resonance imaging at 3.0T. Arthritis Rheum 52(10):3132–3136. doi:10.1002/art.21348

Eckstein F, Lemberger B, Gratzke C, Hudelmaier M, Glaser C, Englmeier KH, Reiser M (2005b) In vivo cartilage deformation after different types of activity and its dependence on physical training status. Ann Rheum Dis 64(2):291–295

Eckstein F, Cicuttini F, Raynauld JP, Waterton JC, Peterfy C (2006) Magnetic resonance imaging (MRI) of articular cartilage in knee osteoarthritis (OA): morphological assessment. Osteoarthritis Cartilage 14(Suppl A):A46–75. doi:10.1016/j.joca.2006.02.026

Eckstein F, Hudelmaier M, Cahue S, Marshall M, Sharma L (2009) Medial-to-lateral ratio of tibiofemoral subchondral bone area is adapted to alignment and mechanical load. Calcif Tissue Int 84(3):186–194. doi:10.1007/s00223-008-9208-4

Erhart-Hledik JC, Favre J, Asay JL, Smith RL, Giori NJ, Mundermann A, Andriacchi TP (2012) A relationship between mechanically-induced changes in serum cartilage oligomeric matrix protein (COMP) and changes in cartilage thickness after 5 years. Osteoarthritis Cartilage 20 (11):1309–1315. doi:10.1016/j.joca.2012.07.018

Evans EB, Eggers GWN, Butler JK, Blumel J (1960) Experimental immobilization and remobilization of rat knee joints. J Bone Joint Surg Am 42:737–758

Farquhar T, Xia Y, Mann K, Bertram J, Burton-Wurster N, Jelinski L, Lust G (1996) Swelling and fibronectin accumulation in articular cartilage explants after cyclical impact. J Orthop Res 14 (3):417–423. doi:10.1002/jor.1100140312

Felson DT, Lawrence RC, Dieppe PA, Hirsch R, Helmick CG, Jordan JM, Kington RS, Lane NE, Nevitt MC, Zhang Y, Sowers M, McAlindon T, Spector TD, Poole AR, Yanovski SZ, Ateshian G, Sharma L, Buckwalter JA, Brandt KD, Fries JF (2000) Osteoarthritis: new insights. Part 1: the disease and its risk factors. Ann Intern Med 133(8):635–646

Freed LE, Langer R, Martin I, Pellis NR, Vunjak-Novakovic G (1997) Tissue engineering of cartilage in space. Proc Natl Acad Sci USA 94(25):13885–13890

Freeman MA, Pinskerova V (2005) The movement of the normal tibio-femoral joint. J Biomech 38 (2):197–208. doi:10.1016/j.jbiomech.2004.02.006

Giannoni P, Siegrist M, Hunziker EB, Wong M (2003) The mechanosensitivity of cartilage oligomeric matrix protein (COMP). Biorheology 40(1–3):101–109

Griffin TM, Guilak F (2005) The role of mechanical loading in the onset and progression of osteoarthritis. Exerc Sport Sci Rev 33(4):195–200

Grimm D, Wehland M, Pietsch J, Aleshcheva G, Wise P, van Loon J, Ulbrich C, Magnusson NE, Infanger M, Bauer J (2014) Growing tissues in real and simulated microgravity: new methods for tissue engineering. Tissue Eng Part B Rev 20(6):555–566. doi:10.1089/ten.TEB.2013.0704

Haapala J, Arokoski JP, Hyttinen MM, Lammi M, Tammi M, Kovanen V, Helminen HJ, Kiviranta I (1999) Remobilization does not fully restore immobilization induced articular cartilage atrophy. Clin Orthop Relat Res 362:218–229

Haapala J, Arokoski J, Pirttimaki J, Lyyra T, Jurvelin J, Tammi M, Helminen HJ, Kiviranta I (2000) Incomplete restoration of immobilization induced softening of young beagle knee articular cartilage after 50-week remobilization. Int J Sports Med 21(1):76–81

Hagiwara Y, Ando A, Chimoto E, Saijo Y, Ohmori-Matsuda K, Itoi E (2009) Changes of articular cartilage after immobilization in a rat knee contracture model. J Orthop Res 27(2):236–242. doi:10.1002/jor.20724

Halasz K, Kassner A, Morgelin M, Heinegard D (2007) COMP acts as a catalyst in collagen fibrillogenesis. J Biol Chem 282(43):31166–31173. doi:10.1074/jbc.M705735200

Hamann N, Zaucke F, Heilig J, Oberlander KD, Brüggemann GP, Niehoff A (2012) Effect of different running modes on the morphological, biochemical, and mechanical properties of articular cartilage. Scand J Med Sci Sports 24(1):179–188. doi:10.1111/j.1600-0838.2012.01513.x

Hamann N, Zaucke F, Dayakli M, Bruggemann GP, Niehoff A (2013) Growth-related structural, biochemical, and mechanical properties of the functional bone-cartilage unit. J Anat 222 (2):248–259

Hedbom E, Antonsson P, Hjerpe A, Aeschlimann D, Paulsson M, Rosa-Pimentel E, Sommarin Y, Wendel M, Oldberg A, Heinegard D (1992) Cartilage matrix proteins. An acidic oligomeric protein (COMP) detected only in cartilage. J Biol Chem 267(9):6132–6136

Heinegard D (2009) Proteoglycans and more—from molecules to biology. Int J Exp Pathol 90 (6):575–586. doi:10.1111/j.1365-2613.2009.00695.x

Herzog W, Federico S (2007) Articular cartilage. In: Nigg BM, Herzog W (eds) Biomechanics of the musculo-skeletal system, 3rd edn. Wiley, West Sussex, pp 95–109

Hinterwimmer S, Krammer M, Krotz M, Glaser C, Baumgart R, Reiser M, Eckstein F (2004) Cartilage atrophy in the knees of patients after seven weeks of partial load bearing. Arthritis Rheum 50(8):2516–2520

Hoson T, Kamisaka S, Masuda Y, Yamashita M, Buchen B (1997) Evaluation of the three-dimensional clinostat as a simulator of weightlessness. Planta 203:S187–S197

Hsu SH, Kuo CC, Yen HJ, Whu SW, Tsai CL (2005) The effect of two different bioreactors on the neocartilage formation in type II collagen modified polyester scaffolds seeded with chondrocytes. Artif Organs 29(6):467–474. doi:10.1111/j.1525-1594.2005.29080.x

Hudelmaier M, Glaser C, Hausschild A, Burgkart R, Eckstein F (2006) Effects of joint unloading and reloading on human cartilage morphology and function, muscle cross-sectional areas, and bone density – a quantitative case report. J Musculoskelet Neuronal Interact 6(3):284–290
Hung CT, Mow VC (2012) Biomechanics of articular cartilage. In: Nordin M, Frankel VH (eds) Basic biomechanics of the musculoskeletal system, 4th edn. Lippincott Williams & Wilkins/A Wolters Kluwer Business, Philadelphia, pp 60–101
Hunter DJ, Altman RD, Cicuttini F, Crema MD, Duryea J, Eckstein F, Guermazi A, Kijowski R, Link TM, Martel-Pelletier J, Miller CG, Mosher TJ, Ochoa-Albiztegui RE, Pelletier JP, Peterfy C, Raynauld JP, Roemer FW, Totterman SM, Gold GE (2015) OARSI clinical trials recommendations: knee imaging in clinical trials in osteoarthritis. Osteoarthritis Cartilage 23 (5):698–715. doi:10.1016/j.joca.2015.03.012
Jeffrey JE, Gregory DW, Aspden RM (1995) Matrix damage and chondrocyte viability following a single impact load on articular cartilage. Arch Biochem Biophys 322(1):87–96. doi:10.1006/abbi.1995.1439
Johnson RB (1998) The bearable lightness of being: bones, muscles, and spaceflight. Anat Rec 253 (1):24–27
Johnson A, Smith R, Saxne T, Hickery M, Heinegard D (2004) Fibronectin fragments cause release and degradation of collagen-binding molecules from equine explant cultures. Osteoarthritis Cartilage 12(2):149–159
Jordan JM (2005) Update on cartilage oligomeric matrix protein as a marker of osteoarthritis. J Rheumatol 32(6):1145–1147
Jurvelin J, Kiviranta I, Tammi M, Helminen JH (1986) Softening of canine articular cartilage after immobilization of the knee joint. Clin Orthop Relat Res 207:246–252
Kahn J, Shwartz Y, Blitz E, Krief S, Sharir A, Breitel DA, Rattenbach R, Relaix F, Maire P, Rountree RB, Kingsley DM, Zelzer E (2009) Muscle contraction is necessary to maintain joint progenitor cell fate. Dev Cell 16(5):734–743. doi:10.1016/j.devcel.2009.04.013
Kakurin LI, Lobachik VI, Mikhailov VM, Senkevich YA (1976) Antiorthostatic hypokinesia as a method of weightlessness simulation. Aviat Space Environ Med 47(10):1083–1086
Kim HJ, Lee YH, Kim CK (2007) Biomarkers of muscle and cartilage damage and inflammation during a 200 km run. Eur J Appl Physiol 99(4):443–447. doi:10.1007/s00421-006-0362-y
Kim HJ, Lee YH, Kim CK (2009) Changes in serum cartilage oligomeric matrix protein (COMP), plasma CPK and plasma hs-CRP in relation to running distance in a marathon (42.195 km) and an ultra-marathon (200 km) race. Eur J Appl Physiol 105(5):765–770. doi:10.1007/s00421-008-0961-x
Kiviranta I, Jurvelin J, Tammi M, Saamanen AM, Helminen HJ (1987) Weight bearing controls glycosaminoglycan concentration and articular cartilage thickness in the knee joints of young beagle dogs. Arthritis Rheum 30(7):801–809
Kiviranta I, Tammi M, Jurvelin J, Saamanen AM, Helminen HJ (1988) Moderate running exercise augments glycosaminoglycans and thickness of articular cartilage in the knee joint of young beagle dogs. J Orthop Res 6(2):188–195
Kiviranta I, Tammi M, Jurvelin J, Arokoski J, Saamanen AM, Helminen HJ (1992) Articular cartilage thickness and glycosaminoglycan distribution in the canine knee joint after strenuous running exercise. Clin Orthop Relat Res 283:302–308
Klement BJ, Spooner BS (1999) Mineralization and growth of cultured embryonic skeletal tissue in microgravity. Bone 24(4):349–359
Kraus VB, Nevitt M, Sandell LJ (2010) Summary of the OA biomarkers workshop 2009—biochemical biomarkers: biology, validation, and clinical studies. Osteoarthritis Cartilage 18 (6):742–745. doi:10.1016/j.joca.2010.02.014
LeBlanc A, Schneider V, Shackelford L, West S, Oganov V, Bakulin A, Voronin L (2000) Bone mineral and lean tissue loss after long duration space flight. J Musculoskelet Neuronal Interact 1(2):157–160
Lelkes G (1958) Experiments in vitro on the role of movement in the development of joints. J Embryol Exp Morphol 6(2):183–186

Leong DJ, Gu XI, Li Y, Lee JY, Laudier DM, Majeska RJ, Schaffler MB, Cardoso L, Sun HB (2010) Matrix metalloproteinase-3 in articular cartilage is upregulated by joint immobilization and suppressed by passive joint motion. Matrix Biol 29(5):420–426. doi:10.1016/j.matbio.2010.02.004

LeRoux MA, Cheung HS, Bau JL, Wang JY, Howell DS, Setton LA (2001) Altered mechanics and histomorphometry of canine tibial cartilage following joint immobilization. Osteoarthritis Cartilage 9(7):633–640

Lindqvist E, Eberhardt K, Bendtzen K, Heinegard D, Saxne T (2005) Prognostic laboratory markers of joint damage in rheumatoid arthritis. Ann Rheum Dis 64(2):196–201. doi:10.1136/ard.2003.019992

Liphardt AM, Mundermann A, Koo S, Backer N, Andriacchi TP, Zange J, Mester J, Heer M (2009) Vibration training intervention to maintain cartilage thickness and serum concentrations of cartilage oligometric matrix protein (COMP) during immobilization. Osteoarthritis Cartilage 17(12):1598–1603

Liphardt AM, Brüggemann GP, Hamann N, Zaucke F, Eckstein F, Bloch W, Mündermann A, Koo S, Mester J, Niehoff A (2015) The effect of immobility and microgravity on cartilage metabolism. Ann Rheum Dis 74(Suppl 2):919

Mankin HJ, Mow VC, Buckwalter JA, Iannotti JP, Ratcliffe A (1999) Articular cartilage structure, composition, and function. In: Buckwalter JA, Einhorn TA, Simon SR (eds) Orthopedic basic science: biology and biomechanics of the musculoskeletal system. American Academy of Orthopaedic Surgeons, Rosemont, pp 440–470

Mann HH, Ozbek S, Engel J, Paulsson M, Wagener R (2004) Interactions between the cartilage oligomeric matrix protein and matrilins. Implications for matrix assembly and the pathogenesis of chondrodysplasias. J Biol Chem 279(24):25294–25298

Mansson B, Carey D, Alini M, Ionescu M, Rosenberg LC, Poole AR, Heinegard D, Saxne T (1995) Cartilage and bone metabolism in rheumatoid arthritis. Differences between rapid and slow progression of disease identified by serum markers of cartilage metabolism. J Clin Invest 95 (3):1071–1077. doi:10.1172/jci117753

Maroudas AI (1976) Balance between swelling pressure and collagen tension in normal and degenerate cartilage. Nature 260(5554):808–809

Monfort J, Garcia-Giralt N, Lopez-Armada MJ, Monllau JC, Bonilla A, Benito P, Blanco FJ (2006) Decreased metalloproteinase production as a response to mechanical pressure in human cartilage: a mechanism for homeostatic regulation. Arthritis Res Ther 8(5):R149

Montufar-Solis D, Duke PJ (1999) Gravitational changes affect tibial growth plates according to Hert's curve. Aviat Space Environ Med 70(3 Pt 1):245–249

Moriyama H, Yoshimura O, Kawamata S, Takayanagi K, Kurose T, Kubota A, Hosoda M, Tobimatsu Y (2008) Alteration in articular cartilage of rat knee joints after spinal cord injury. Osteoarthritis Cartilage 16(3):392–398. doi:10.1016/j.joca.2007.07.002

Morrison CJ, Butler GS, Rodriguez D, Overall CM (2009) Matrix metalloproteinase proteomics: substrates, targets, and therapy. Curr Opin Cell Biol 21(5):645–653

Mosher TJ, Dardzinski BJ (2004) Cartilage MRI T2 relaxation time mapping: overview and applications. Semin Musculoskelet Radiol 8(4):355–368

Moss ML, Moss-Salentijn L (1983) Vertebrate cartilages. In: Hall BK (ed) Cartilage – structure, function and biochemistry, vol 1. Academic, New York, pp 1–30

Muhlbauer R, Lukasz TS, Faber TS, Stammberger T, Eckstein F (2000) Comparison of knee joint cartilage thickness in triathletes and physically inactive volunteers based on magnetic resonance imaging and three-dimensional analysis. Am J Sports Med 28(4):541–546

Mundermann A, Dyrby CO, Andriacchi TP, King KB (2005) Serum concentration of cartilage oligomeric matrix protein (COMP) is sensitive to physiological cyclic loading in healthy adults. Osteoarthritis Cartilage 13(1):34–38. doi:10.1016/j.joca.2004.09.007

Mundermann A, King KB, Smith RL, Andriacchi TP (2009) Change in serum COMP concentration due to ambulatory load is not related to knee OA Status. J Orthop Res 27(11):1408–1413

Murray PD, Drachman DB (1969) The role of movement in the development of joints and related structures: the head and neck in the chick embryo. J Embryol Exp Morphol 22(3):349–371
Neidhart M, Hauser N, Paulsson M, DiCesare PE, Michel BA, Hauselmann HJ (1997) Small fragments of cartilage oligomeric matrix protein in synovial fluid and serum as markers for cartilage degradation. Br J Rheumatol 36(11):1151–1160
Neidhart M, Muller-Ladner U, Frey W, Bosserhoff AK, Colombani PC, Frey-Rindova P, Hummel KM, Gay RE, Hauselmann H, Gay S (2000) Increased serum levels of non-collagenous matrix proteins (cartilage oligomeric matrix protein and melanoma inhibitory activity) in marathon runners. Osteoarthritis Cartilage 8(3):222–229. doi:10.1053/joca.1999.0293
Newman B, Wallis GA (2002) Is osteoarthritis a genetic disease? Clin Invest Med 25(4):139–149
Niehoff A, Offermann M, Dargel J, Schmidt A, Brüggemann GP, Bloch W (2008) Dynamic and static mechanical compression affects Akt phosphorylation in porcine patellofemoral joint cartilage. J Orthop Res 26(5):616–623. doi:10.1002/jor.20542
Niehoff A, Kersting UG, Helling S, Dargel J, Maurer J, Thevis M, Bruggemann GP (2010) Different mechanical loading protocols influence serum cartilage oligomeric matrix protein levels in young healthy humans. Eur J Appl Physiol 110(3):651–657
Niehoff A, Muller M, Bruggemann L, Savage T, Zaucke F, Eckstein F, Muller-Lung U, Bruggemann GP (2011) Deformational behaviour of knee cartilage and changes in serum cartilage oligomeric matrix protein (COMP) after running and drop landing. Osteoarthritis Cartilage 19(8):1003–1010
Nowlan NC, Murphy P, Prendergast PJ (2007) Mechanobiology of embryonic limb development. Ann N Y Acad Sci 1101:389–411. doi:10.1196/annals.1389.003
O'Rahilly R, Gardner E (1978) The embryology of movable joints. In: Sokoloff L (ed) The joints and synovial fluid, vol 1. Academic, New York, pp 49–103
Okada Y, Shinmei M, Tanaka O, Naka K, Kimura A, Nakanishi I, Bayliss MT, Iwata K, Nagase H (1992) Localization of matrix metalloproteinase 3 (stromelysin) in osteoarthritic cartilage and synovium. Lab Invest 66(6):680–690
Owman H, Tiderius CJ, Ericsson YB, Dahlberg LE (2014) Long-term effect of removal of knee joint loading on cartilage quality evaluated by delayed gadolinium-enhanced magnetic resonance imaging of cartilage. Osteoarthritis Cartilage 22(7):928–932. doi:10.1016/j.joca.2014.04.021
Parkkinen JJ, Lammi MJ, Helminen HJ, Tammi M (1992) Local stimulation of proteoglycan synthesis in articular cartilage explants by dynamic compression in vitro. J Orthop Res 10(5):610–620
Pavy-Le Traon A, Heer M, Narici MV, Rittweger J, Vernikos J (2007) From space to earth: advances in human physiology from 20 years of bed rest studies (1986–2006). Eur J Appl Physiol 101(2):143–194
Pearle AD, Warren RF, Rodeo SA (2005) Basic science of articular cartilage and osteoarthritis. Clin Sports Med 24(1):1–12. doi:10.1016/j.csm.2004.08.007
Peterfy CG, Schneider E, Nevitt M (2008) The osteoarthritis initiative: report on the design rationale for the magnetic resonance imaging protocol for the knee. Osteoarthritis Cartilage 16(12):1433–1441
Petersson IF, Boegard T, Dahlstrom J, Svensson B, Heinegard D, Saxne T (1998) Bone scan and serum markers of bone and cartilage in patients with knee pain and osteoarthritis. Osteoarthritis Cartilage 6(1):33–39. doi:10.1053/joca.1997.0090
Pitsillides AA (2006) Early effects of embryonic movement: 'a shot out of the dark'. J Anat 208(4):417–431. doi:10.1111/j.1469-7580.2006.00556.x
Rehan Youssef A, Longino D, Seerattan R, Leonard T, Herzog W (2009) Muscle weakness causes joint degeneration in rabbits. Osteoarthritis Cartilage 17(9):1228–1235. doi:10.1016/j.joca.2009.03.017
Rosenberg K, Olsson H, Morgelin M, Heinegard D (1998) Cartilage oligomeric matrix protein shows high affinity zinc-dependent interaction with triple helical collagen. J Biol Chem 273(32):20397–20403

Roth JH, Mendenhall HV, McPherson GK (1988) The effect of immobilization on goat knees following reconstruction of the anterior cruciate ligament. Clin Orthop Relat Res 229:278–282

Sah RL, Kim YJ, Doong JY, Grodzinsky AJ, Plaas AH, Sandy JD (1989) Biosynthetic response of cartilage explants to dynamic compression. J Orthop Res 7(5):619–636

Sauerland K, Raiss RX, Steinmeyer J (2003) Proteoglycan metabolism and viability of articular cartilage explants as modulated by the frequency of intermittent loading. Osteoarthr Cartil 11 (5):343–350. doi:10.1016/S1063-4584(03)00007-4

Saxne T, Heinegard D (1992) Cartilage oligomeric matrix protein: a novel marker of cartilage turnover detectable in synovial fluid and blood. Br J Rheumatol 31(9):583–591

Scotece M, Mobasheri A (2015) Leptin in osteoarthritis: Focus on articular cartilage and chondrocytes. Life Sci 140:75–78. doi:10.1016/j.lfs.2015.05.025

Setton LA, Mow VC, Muller FJ, Pita JC, Howell DS (1997) Mechanical behavior and biochemical composition of canine knee cartilage following periods of joint disuse and disuse with remobilization. Osteoarthritis Cartilage 5(1):1–16

Shefelbine SJ, Carter DR (2004) Mechanobiological predictions of growth front morphology in developmental hip dysplasia. J Orthop Res 22(2):346–352. doi:10.1016/j.orthres.2003.08.004

Shwartz Y, Blitz E, Zelzer E (2013) One load to rule them all: mechanical control of the musculoskeletal system in development and aging. Differentiation 86(3):104–111. doi:10.1016/j.diff.2013.07.003

Smith RL, Thomas KD, Schurman DJ, Carter DR, Wong M, van der Meulen MC (1992) Rabbit knee immobilization: bone remodeling precedes cartilage degradation. J Orthop Res 10 (1):88–95

Steinmeyer J, Ackermann B, Raiss RX (1997) Intermittent cyclic loading of cartilage explants modulates fibronectin metabolism. Osteoarthr Cartil 5(5):331–341. doi:10.1016/S1063-4584 (97)80037-4

Stokes IA, Iatridis JC (2004) Mechanical conditions that accelerate intervertebral disc degeneration: overload versus immobilization. Spine (Phila Pa 1976) 29(23):2724–2732

Svoboda SJ, Harvey TM, Owens BD, Brechue WF, Tarwater PM, Cameron KL (2013) Changes in serum biomarkers of cartilage turnover after anterior cruciate ligament injury. Am J Sports Med 41(9):2108–2116. doi:10.1177/0363546513494180

Tetlow LC, Woolley DE (2001) Expression of vitamin D receptors and matrix metalloproteinases in osteoarthritic cartilage and human articular chondrocytes in vitro. Osteoarthritis Cartilage 9 (5):423–431. doi:10.1053/joca.2000.0408

Tetlow LC, Adlam DJ, Woolley DE (2001) Matrix metalloproteinase and proinflammatory cytokine production by chondrocytes of human osteoarthritic cartilage: associations with degenerative changes. Arthritis Rheum 44(3):585–594

Thambyah A (2007) Contact stresses in both compartments of the tibiofemoral joint are similar even when larger forces are applied to the medial compartment. Knee 14(4):336–338. doi:10.1016/j.knee.2007.05.002

Thompson RC Jr, Oegema TR Jr, Lewis JL, Wallace L (1991) Osteoarthrotic changes after acute transarticular load. An animal model. J Bone Joint Surg Am 73(7):990–1001

Tomiya M, Fujikawa K, Ichimura S, Kikuchi T, Yoshihara Y, Nemoto K (2009) Skeletal unloading induces a full-thickness patellar cartilage defect with increase of urinary collagen II CTx degradation marker in growing rats. Bone 44(2):295–305. doi:10.1016/j.bone.2008.10.038

Ulbrich C, Westphal K, Pietsch J, Winkler HD, Leder A, Bauer J, Kossmehl P, Grosse J, Schoenberger J, Infanger M, Egli M, Grimm D (2010) Characterization of human chondrocytes exposed to simulated microgravity. Cell Physiol Biochem 25(4–5):551–560. doi:10.1159/000303059

Vanwanseele B, Eckstein F, Knecht H, Stussi E, Spaepen A (2002) Knee cartilage of spinal cord-injured patients displays progressive thinning in the absence of normal joint loading and movement. Arthritis Rheum 46(8):2073–2078

Vanwanseele B, Eckstein F, Knecht H, Spaepen A, Stussi E (2003) Longitudinal analysis of cartilage atrophy in the knees of patients with spinal cord injury. Arthritis Rheum 48 (12):3377–3381
Warwick R, Williams P (1973) Gray's Anatomy. W.B. Saunders, Philadelphia
Wong M, Carter DR (1988) Mechanical stress and morphogenetic endochondral ossification of the sternum. J Bone Joint Surg Am 70(7):992–1000
Wong M, Wuethrich P, Eggli P, Hunziker E (1996) Zone-specific cell biosynthetic activity in mature bovine articular cartilage: a new method using confocal microscopic stereology and quantitative autoradiography. J Orthop Res 14(3):424–432. doi:10.1002/jor.1100140313
Wong M, Wuethrich P, Buschmann MD, Eggli P, Hunziker E (1997) Chondrocyte biosynthesis correlates with local tissue strain in statically compressed adult articular cartilage. J Orthop Res 15(2):189–196. doi:10.1002/jor.1100150206
Wong M, Siegrist M, Cao X (1999) Cyclic compression of articular cartilage explants is associated with progressive consolidation and altered expression pattern of extracellular matrix proteins. Matrix Biol 18(4):391–399
Yong VW, Agrawal SM, Stirling DP (2007) Targeting MMPs in acute and chronic neurological conditions. Neurotherapeutics 4(4):580–589

Chapter 3
Influence of Weightlessness on Aerobic Capacity, Cardiac Output and Oxygen Uptake Kinetics

U. Hoffmann, A.D. Moore, J. Koschate, and U. Drescher

Abstract The exposure to weightlessness can have an impact on aerobic capacity as a result of changes in the cardiorespiratory and musculoskeletal systems. As a consequence, astronauts' work capacities might be changed which would affect activities during the Space missions and after return to Earth or other environments with gravity, i.e. Mars or Moon. This chapter will give an overview about results from studies using cardiopulmonary exercise testing (CPET). This method allows to monitor astronauts' fitness non-invasively and is, therefore, qualified for inflight measurements. Results from early Space flight until now will be compared with results from bedrest studies. Predominantly, the focus lies on peak oxygen uptake, heart rate and oxygen uptake kinetics. Since the specific methods and problems of CPET are related to the concepts of respiratory gas measurement, aspects of hardware, exercise protocols and data analysis will be discussed.

Keywords Weightlessness • Microgravity • Space • peak oxygen uptake • heart rate kinetics • oxygen uptake kinetics • bed rest

3.1 Introduction

The ability to sustain workloads with high portions of aerobic metabolism is critical in the maintenance of physical work capacity, both on Earth and in Space. Cardiopulmonary exercise testing (CPET) is an excellent, non-invasive tool to quantify

U. Hoffmann (✉) • J. Koschate • U. Drescher
Institute of Physiology and Anatomy, German Sport University Cologne, Am Sportpark Müngersdorf 6, 50933 Cologne, Germany
e-mail: u.hoffmann@dshs-koeln.de

A.D. Moore
Institute of Physiology and Anatomy, German Sport University Cologne, Am Sportpark Müngersdorf 6, 50933 Cologne, Germany

Department of Health and Kinesiology, Lamar University, P.O. Box 10039, Beaumont, TX 77710, USA

S. Schneider (ed.), *Exercise in Space*, SpringerBriefs in Space Life Sciences,
DOI 10.1007/978-3-319-29571-8_3

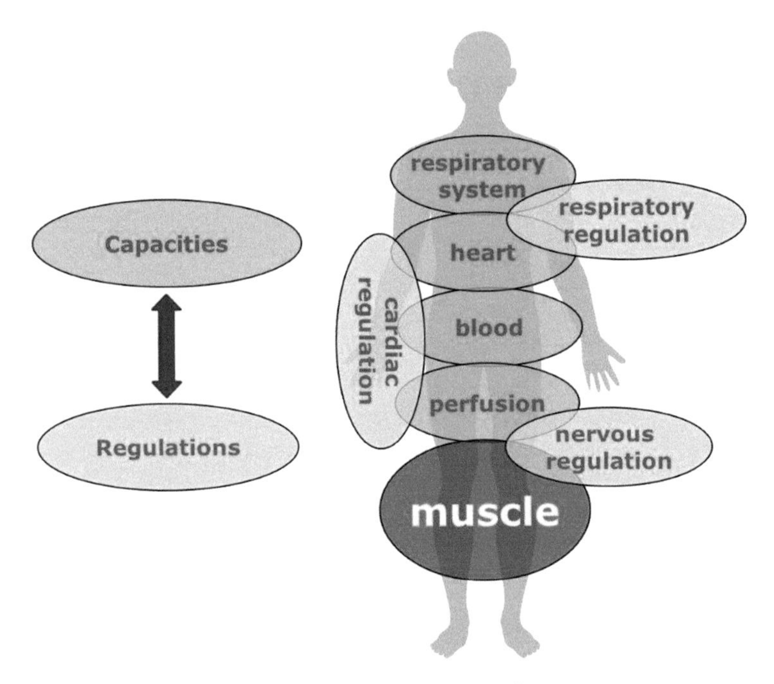

Fig. 3.1 Processes which may limit muscular aerobic metabolism

both aerobic capacity and aspects of cardiorespiratory regulation related to the capacity to perform physical work. The purpose of this chapter is to summarise the results of CPET related to spaceflight and the methodological aspects for these special conditions.

During exercise testing, measurement of pulmonary O_2 uptake ($\dot{V}O_2$pulm) and other cardiorespiratory parameters [e.g. heart rate (HR) and ventilation ($\dot{V}E$)] allows assessment of both systemic and local muscular responses which may limit aerobic metabolism and O_2 transport (Fig. 3.1). Therefore, the weightlessness-induced acute and long-term adaptive effects on both the cardiorespiratory and the musculoskeletal system can be investigated by CPET.

However, spaceflight causes not only weightlessness-related hypokinesia but also changes of hydrostatic and compression factors (Greenleaf 1984) which contribute to adaptations. Furthermore, environmental in-flight conditions, such as confinement, ambient pressure and temperature as well as partial pressures of oxygen (PO_2) and carbon dioxide (PCO_2) (Moore et al. 2010) which may differ from standard environmental conditions on Earth, may contribute to cardiopulmonary deconditioning. These factors, in addition to the effects of physical training as countermeasures and other strenuous exercises, e.g. extra vehicular activities (EVA), complicate the interpretation of CPET results due to overlapping and confounding stimuli for adaptation to weightlessness.

With an adequate CPET, aerobic metabolic responses of muscles and the O_2 transport during exercise can be monitored during long-term Space missions. The purpose of this monitoring is to track the physiological changes related to aerobic metabolism in order to predict work capacity, both while in Space and after the return to gravity.

It is challenging to optimise CPET and the interpretations of results to the specific environment and requirements during spaceflight missions. The following factors contribute to both the development of CPET procedures and interpretation of CPET results: (a) testing procedure, i.e. instrumentation for gas exchange measurements, ergometer and work rate (WR) protocol; (b) environmental conditions, i.e. ambient pressure, O_2 and CO_2 concentrations; and (c) influences of countermeasures and extraordinary activities during the mission, e.g. physical training, exercise tests and EVA. These factors are explored in this article. Trappe et al. (2005) primarily examined the effects of spaceflight on the muscle in their study and stated, "...that the data ... do not describe the effects of weightlessness alone ... on physical working capacity, rather the effects of living in Space", this statement holds true for all data collected from Space missions.

3.2 Study of Aerobic Capacities Versus Study of Cardiopulmonary Regulation in Spaceflight

The research results discussed in this section must be examined carefully with respect to differences in methodology between investigations (see Sect. 3.3 for details). This is particularly important when studies performed on Earth, such as the often-used bedrest analogue of spaceflight, are compared to those conducted in, or following, actual microgravity exposure. Another important factor that should be recognised is that the fitness levels and other individual characteristics, such as age, of astronauts and subjects of ground-based studies oftentimes differ—with a bias of the ground studies typically using a younger cohort of subjects. An overview for selected studies is given in Table 3.1.

Very early in the history of manned spaceflights, it became obvious that cardiovascular adaptations occurred which likely limited exercise capacity after return to Earth. Within the US Space programme, the first indication of cardiovascular dysfunction was orthostatic intolerance following a 34 h Mercury Program flight (Moore et al. 2010). Following the Gemini Program flights, which were ~3–14 days in duration, it became obvious that not only exercise capacity but also the involved physiological regulation of the responses to exercise themselves limited aerobic metabolism. Two aspects regarding the limitation of aerobic metabolism induced by spaceflight must be considered. The first aspect relates to changes that occur after the entry in weightlessness during the sojourn in Space, and the second are limitations that are apparent after return to gravity.

The regulatory requirements with altering the rate of aerobic metabolism after changes in WR are regulation of O_2 transport to the exercising muscles by cardiac output ($\dot{Q}$) and the synchronisation of local blood perfusion with muscular O_2 demands. This involves adaptations of HR and stroke volume (SV) as well as adequate peripheral vascular regulation.

Table 3.1 Comparison of different $\dot{V}O_2$ peak, $\dot{V}O_2$/HR kinetics results in the context of Space flight missions and bed rest studies with regard to methodological aspects

References	Number of subjects	Condition	Duration	Day of measurement	Posture	Spirometry	Test protocol	Countermeasures	Results
Space missions—$\dot{V}O_2$peak									
Moore et al. (2014)	14	Ground		L−90 R+1, R+10, R+30	Upright seated	Mixing box	3-min stages—25 %, 50 % and 75 % (reference: preflight $\dot{V}O_2$peak test) every min 25 W increments until symptom-limited maximum	Y	In-flight lower $\dot{V}O_2$peak than preflight with increase over the mission days, decrease after landing with recovery within 30 days
		Space	180 (mean)	L+15, L+45, etc.	"Upright" waist belt, side-mounted hand grips at waist level				
Trappe et al. (2005)	4	Ground		L−60, L−30, L−15 R+1, R+4/5, R+8	Semi-recumbent	Mixing box	3-min stages—50 W, 100 W, 150 W, 175 W every 2 min 25 W increments until exhaustion	Y	In-flight lower $\dot{V}O_2$peak than preflight with decrease over the mission days, lower but not significant $\dot{V}O_2$ after the flight
		Space	17	L+2, L+8, L +13	Semi-recumbent		3-min stages—50 W, 100 W, 150 W, 175 W, every 2 min 25 W increments until 85 % of the peak WR (reference: preflight $\dot{V}O_2$peak test)		

Levine et al. (1996)	6	Ground		L−120, L−90, L−60, L−15 R+0, R+1/2, R+6/7	Upright	Mixing box	5-min warmup, 25 min, 30 %; 15 min, 60 % (reference: initial $\dot{V}O_2$peak test) followed by variants SLS-1: 3 min 90 %, at least 2 min, 100 %, SLS-2: increments of 10 W–25 W every minute until exhaustion	Y	No difference between preflight and in-flight $\dot{V}O_2$peak, $\dot{V}O_2$peak significantly declined on R+0 and R+1
		Space	9/14	L+5−8	"Upright" Restrained with pads fixed at shoulder level				
Stegemann et al. (1997)	4	Ground	10	L−137, L−13 R+2, R+15	Upright	Breath-by-breath	3 min—20 W, 450 s—20 W–80 W randomised changes, 4.5 min—80 W, increments of 10 W—30 s until subjective exhaustion	N	Lower, but non-significant $\dot{V}O_2$peak after the flight, significantly lower WR at aerobic-anaerobic threshold on R+2
Bedrest—$\dot{V}O_2$peak									
Convertino et al. (1982)	12	Bedrest	10	BR−1 R+0	Supine/upright	Mixing box	3 min—20 %, 45 %, 70 % and 100 % (reference: initial $\dot{V}O_2$peak test)	N	Significant decrease in $\dot{V}O_2$peak after BR in supine position

(continued)

Table 3.1 (continued)

References	Number of subjects	Condition	Duration	Day of measurement	Posture	Spirometry	Test protocol	Countermeasures	Results
Ferretti et al. (1997)	7	Bedrest	42	Pre-R+4	Upright	1 min Douglas bag collection	5 min—100 W (before BR)/50 W (after BR), every 5 min 50 W increments/25 W increment at expected individual maximum WR	N	Significant $\dot{V}O_2$peak reduction after BR
Trappe et al. (2005)	8	Bedrest	17	BR−12, BR−7 BR+2, BR +8, BR+13, R+3, R+7	Supine	Mixing box	3-min stages—50 W, 100 W, 150 W, 200 W, every 2 min 25 W increments until exhaustion	Y	Significant steady VO_2peak decrease during BR and on R+3
Bedrest/Space missions—$\dot{V}O_2$/HR kinetics									
Convertino et al. (1984)	5	Bedrest	7	BR−12/−2 R+0	Supine/ upright	Mixing box	5 min rest, 5 min of exercise at 115 W, 10 min—rest	N	Slower $\dot{V}O_2$pulm kinetics after BR
Stegemann et al. (1985)	6	Bedrest	7	BR−x R+1, R+3, R+5	Upright	Breath by breath	15 min—20 W, 80 W, 20 W (warmup); 450 s, 20 W, 80 W randomised changes	N	Slower $\dot{V}O_2$pulm kinetics on R+1/R+3

Stegemann et al. (1997)	4	Ground		L−137, L−13 R+2, R+15	Upright	Breath-by-breath	3 min, 20 W; 450 s—20 W, 80 W randomised changes, 4.5 min—80 W	N	Non-significantly slower $\dot{V}O_2$pulm kinetics
		Space	10	L+1, L+3, L +7, L+8	"Upright", strapped on the saddle				
Hoffmann et al. (2016)	10	Ground	153 (mean)	L−236, L−73 R+6, R+21	Upright	Breath-by-breath	200 s—30 W, 300 s—30 W—80 W randomised changes, 200 s—80 W	Y	Significantly slower $\dot{V}O_2$musc kinetics on R+6 and R+21

To some extent, the delayed response of O_2 transport to exercising muscle can be compensated by higher differences of arteriovenous O_2 concentration (a-vDO_2) at muscular site (Wagner 1995). The theoretical a-vDO_2 limit is governed by the O_2 capacity of blood, but when venous PO_2 falls below 5.3 kPa (40 mmHg), diffusive transport between capillaries and mitochondria will delay muscular $\dot{V}O_2$ ($\dot{V}O_2$musc) kinetics (Hughson 2009). Thus, a portion of aerobic metabolism is shifted at least transiently to the anaerobic metabolism, which may lead to an early fatigue, even though the capacities of the physiological subsystems, e.g. $\dot{Q}$ are not exhausted. Similar effects on metabolism can be expected if O_2 diffusion in the lungs, $\dot{Q}$ or local blood flow reaches their exercise-induced maximal capacities. For unspecific workloads with involvement of large muscle groups with non-cyclical movements, such as EVA in Space, the measurement of systemic (whole body) responses specific to the activity as measure for work capacity is most appropriate, although obtaining these measurements can be difficult. The specificity of testing is also important with regard to measurements taken during work tasks on ground (Barstow and Scheuermann 2005). The working limits of repetitive activities with specific muscle groups tend to be limited more by muscular O_2 kinetics than on whole body systemic responses. Ideally, monitoring of both the whole body and local O_2 kinetics is ideal, that is, measurement of capacity of the individual to perform work as well as examination of regulatory mechanisms—including the synchronisation of the transients of O_2 delivery and utilisation with muscular demands. This is relevant to the evaluation of exercise countermeasures, the proper prescription of the countermeasures, and to gain an understanding of the adaptations to weightlessness during Space missions on the physiological subsystems.

A thorough analysis of typical activity demands the aim of countermeasures, and monitoring has to be properly considered. The dimensions to be considered are illustrated in Fig. 3.2. Consideration of these factors allows a selection of

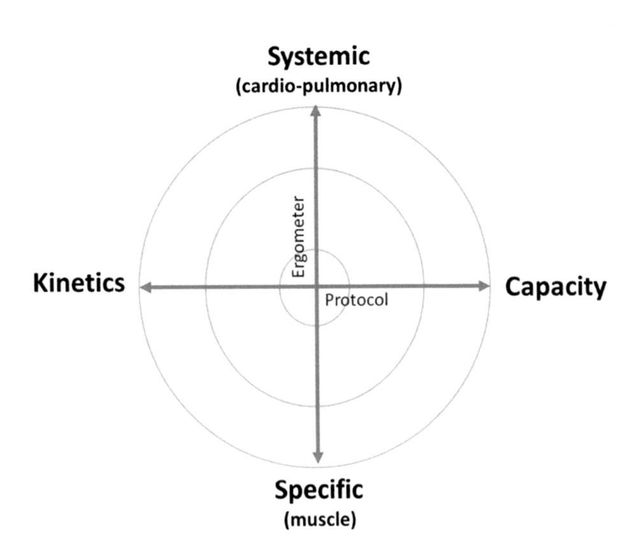

Fig. 3.2 Dimensions to determine the characteristics of exercise demands

adequate CPET design. For example, for astronauts in Space, the target should be close to the centre of both the systemic/specific dimension and to the kinetics/capacity dimension. This will allow evaluation of O_2 transport capacities, aerobic capacities of muscles used for physical work, as well as the related kinetics for all involved subsystems. The ongoing challenge is to design a test which will predict performance ability in Space and following return to gravity on Earth, Mars or the Moon from the characteristics measured in weightlessness by the chosen CPET design.

3.2.1 *Aerobic Capacity and Weightlessness*

3.2.1.1 Results from In-Flight Measurements

Recently Moore et al. (2014) published the first study which tracked the peak O_2 uptake ($\dot{V}O_2$peak) for an investigation period of up to 180 days on the International Space Station (ISS) and the recovery after landing. $\dot{V}O_2$peak, determined by graded exercise tests with measurements of pulmonary gas exchange in combination with HR up to subjective exhaustion, also known as symptom-limited maximum, was used to indicate maximal aerobic capacity. Alternatively, the term maximal O_2 uptake ($\dot{V}O_2$max) is used when several criteria at termination of exercise are met (Midgley et al. 2007). Some of the criteria, such as verifying $\dot{V}O_2$max by application of a separate supramaximal workload in a second test, are problematic due to the time constraints for all biomedical investigations conducted on ISS. In this article, $\dot{V}O_2$peak will be used to designate systemic aerobic capacity.

The initial test of the subjects of the Moore et al. (2014) study consisted of three continuous 3-min stages (either at 50, 75, 100 W or 50, 100, 150 W, depending on the astronauts' body weight and self-report of activity levels), followed by 1-min stepwise increments of 25 W to volitional fatigue. All subsequent testings of the study used a cycle-ergometer-based graded exercise test, consisting of three 5-min levels designed to elicit 25, 50 and 75 % of preflight $\dot{V}O_2$peak, followed by 25 W increases every minute until symptom-limited maximum was attained. This test protocol deviates from those commonly used for patient or athlete examinations with standardised increments, i.e. amplitude and duration, for the whole test (Albouaini et al. 2007).

During spaceflight Moore et al. (2014) observed that the lowest value of both aerobic capacity and peak attained power of the subjects occurred in the first mission days with a slow, but significant increase during the following period (Fig. 3.3). This suggests a loss in $\dot{V}O_2$peak after the onset of weightlessness. With regard to the expected time course of adaptations, systemic adaptations (e.g. plasma volume loss) occurring during the first 2 weeks of exposure are likely more contributing to the decline in $\dot{V}O_2$peak than muscular adaptations, such as decreased intermuscular enzymes involved in aerobic metabolism (Chi et al. 1983).

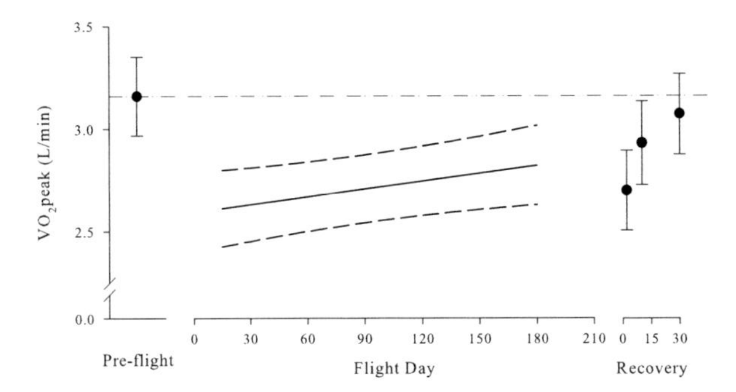

Fig. 3.3 The effect of long-duration spaceflight on absolute peak aerobic capacity ($\dot{V}O_2$peak) during and following ISS missions. The *horizontal broken line* is a reference drawn at the preflight mean. The *solid line ± dashes* during flight is the mixed-modelled linear regression predicted mean response ± 95 % confidence interval (CI) for an average-sized subject (77.2 kg). The values preflight and during recovery are also predicted means ± 95 % CI (from Moore et al. 2014)

Trappe et al. (2005) observed signs of reductions in $\dot{V}O_2$pulm with increasing HR at 85 % of maximal WR (as measured preflight) during three test sessions conducted during the first 13 days of a shuttle flight. Since the peak HRs observed in Space, particularly in the latter days of testing, were close to individual maximal HR measured on ground, this $\dot{V}O_2$pulm was interpreted as strong evidence that $\dot{V}O_2$peak decreased over this period. This is in line with $\dot{V}O_2$peak reduction after 15 days reported by Moore et al. (2014). Both studies, Moore et al. and Trappe et al. (2005), were performed with similar instrumentation and similar exercise protocols, although Trappe et al. (2005) applied a semi-recumbent position, whereas Moore et al. (2014) applied a position similar to upright sitting in gravity. Little information is available about the countermeasures and other physical activities which were conducted during these studies, and the frequency of testing in the study performed by Trappe et al. (2005) may have provided a training stimulus to their subjects. In addition, it was reported in the Moore et al. (2014) study, that the time demands of an ISS crew transition are heavy, and oftentimes exercise periods early during flight are not performed by transitioning crew members.

3.2.1.2 Changes from Gravity to a Weightless Environment

The direct comparison with preflight data in the two studies reported above supports the assumption that the change from gravity to weightlessness reduces $\dot{V}O_2$peak. As written above, these are likely due to alterations occurring in the cardiovascular system and not at the muscle level, even if some objections with regard to the comparability of ground versus Space exercise stimuli may exist. However, in contrast to the results mentioned above, this hypothesis was not supported by the earlier data of Levine et al. (1996) who did not find $\dot{V}O_2$peak reductions in Space after 5–8 flight days. Moore et al. (2010) speculated that this could be explained by subjects preflight $\dot{V}O_2$peak which was lower in the study of Levine et al. (1996) compared to the others, and different hardware for respiratory gas analysis and countermeasures during flight. $\dot{V}O_2$peak was not measured during the NASA Skylab missions conducted in the 1970s (Michel et al. 1977); however, submaximal data were collected during tests using WR loads and progressions similar to the first three stages of the Moore et al. (2014) study. The finding of particular interest in the Skylab M-171 experiment was that exercise HR was similar when preflight and in-flight data were compared. However, there are confounding issues when trying to compare the Skylab results to either ISS or Space Shuttle-based studies. Primarily, the ambient conditions on Skylab were 34.5 kPa cabin pressure and 72 % O_2 concentration.

As an intermediate summary, it can be stated that an early reduction in $\dot{V}O_2$peak is likely related to reduced capacities in the cardiovascular system. For the practical work and exercise demands in Space, this reduction must be considered when planning and conducting activities in Space. Exercise countermeasures as performed in ISS missions did not seem to prevent this early decline, although it is uncertain how often exercise was conducted early in-flight.

To gain a better understanding of simulated weightlessness effects on $\dot{V}O_2$peak, numerous studies can be found in literature using the ground-based analogues for weightlessness, such as neck-out immersion, dry immersion, 6° head-down tilt (HDT), bedrest or combinations. These methods are intended to create a similar hydrostatic pressure gradient to that which occurs during spaceflight and, therefore, produce similar blood and body fluid redistributions like those that happen in weightlessness (Greenleaf 1984). The first, acute effect is related to the blood volume shift from lower body in cranial direction. Therefore, those data are also helpful in the examination of the effects that ensue shortly after posture changes occur and the full onset of diuresis which follow the postural change. Leyk et al. (1994) observed only non-significant decreases in $\dot{V}O_2$peak 20 min after changing to supine position. Convertino et al. (1982) also reported no significant differences between upright and supine $\dot{V}O_2$peak. However, the diuresis effects with haematological changes need longer response time and are fully effective after some hours following posture change (Norsk et al. 1993). This is in line with data from Trappe et al. (2005) who examined subjects after 2 days of bedrest and found a reduced $\dot{V}O_2$peak compared to starting values in supine position.

Most common for prolonged ground-based investigations are HDT bedrest studies. Pavy-Le Traon et al. (2007) reviewed the different aspects of comparability of HDT bedrest and weightlessness. Even though during bedrest a gravitation force is still present, and living in Space does not involve immobilisation similar to that of bedrest, HDT seems a fair simulation of certain aspects of microgravity exposure, particularly those involving the bone, muscle and cardiovascular deconditioning. Haematological effects (Greenleaf 1984) and changes in stroke volume explain at least some losses in $\dot{V}O_2$peak that occur during bedrest deconditioning (Wagner 1995). Bedrest duration, as reported in a review paper by Convertino (1997), is an important contributing factor in the magnitude of reductions in $\dot{V}O_2$peak. This paper summarised bedrest studies of up to 30 days in a meta-analysis and reported a decrease of 0.9 % per day in $\dot{V}O_2$peak. This matches with values reported by Moore et al. (2014) on spaceflight participants. With prolongation of the bedrest duration, the effects that occur after the initial changes in plasma volume distribution assume greater importance to contribute to detraining (Fortney et al. 1991). This raises one point with regard to the ISS paper of Moore et al. (2014). The astronauts involved in the study typically showed their lowest $\dot{V}O_2$peak sometime within the first 45 days of flight, often at day 14. The data set is too small to draw a firm conclusion, but it is likely that the ISS countermeasures are largely attenuating the effects of spaceflight on the cardiopulmonary changes contributing to aerobic capacity loss that are occurring after the initial adjustment to gravity.

Results from studies of detraining effects report significant effects on plasma volume and red cell mass (Mujika and Padilla 2001), a loss in cardiac dimensions and myocardial mass (Mujika and Padilla 2001; Perhonen et al. 2001) and reductions in enzyme concentrations of the exercising muscles (Klausen et al. 1981). Ferretti et al. (1997) identified O_2 transport capacity as the main limiting factor (>70 %) for $\dot{V}O_2$peak. However, the existing $\dot{V}O_2$peak data collected during spaceflight give little hints about the limiting subsystems and potential limitations by slowed or insufficient synchronisation of kinetics of the subsystems.

3.2.1.3 The Shift from Weightlessness Back to a Gravitational Environment

Other issues involving aerobic capacity and spaceflight are the changes observed with return to gravity. Several pre- to postflight comparisons are available and all are in general agreement. $\dot{V}O_2$peak is reduced during tests which are conduced either immediately or shortly after landing (Moore et al. 2001, 2014; Trappe et al. 2005; Stegemann et al. 1997; Levine et al. 1996). Moreover, data from other exercise tests with no direct $\dot{V}O_2$peak estimation support these findings as summarised by Moore et al. (2010). The $\dot{V}O_2$peak decrease seems independent of flight duration. For mission lasting less than 10 days, exercise countermeasures do not prevent this decrease. From data collected before and after the D-2 mission (STS-55) in 1993, Stegemann et al. (1997) determined that subjects who performed no exercise during flight experienced a decline in $\dot{V}O_2$peak similar to those who had performed countermeasures. $\dot{V}O_2$peak reductions were related to hypovolemia and loss of vascular stiffness (Tuday et al. 2007). $\dot{V}O_2$peak data from bedrest studies support this hypothesis by fast recovery within a few days measured in both the upright and supine positions (Trappe et al. 2005; Convertino et al. 1982).

In addition, Stegemann et al. (1997) analysed the incremental test for the anaerobic threshold by the V-slope method (Beaver et al. 1986) and found a significant decrease on R+2. This can be regarded as indication for slowed

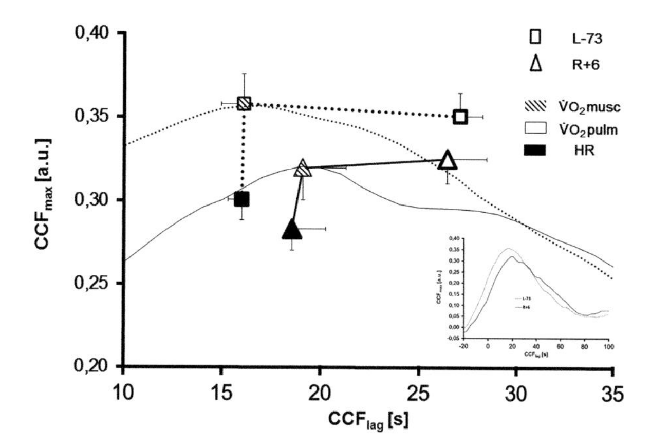

Fig. 3.4 Kinetics responses pre- and post-ISS mission (L-73, R+6) as described by Hoffmann et al. (2016) by time series analysis of PRBS WR responses. The main graph represents maxima and related lags of HR, $\dot{V}O_2$pulm and derived $\dot{V}O_2$musc. For an over-all orientation, the integrated (*lower right*) shows the complete $\dot{V}O_2$musc kinetics on the 2 days of examination

$\dot{V}O_2$musc kinetics, but might also be a change provoked by reduced plasma volume.

The results from Moore et al. (2014) demonstrate that exercise countermeasures and adequate rehabilitation during recovery can compensate for these effects within 30 days after landing; however, little information are available regarding potential changes and recovery of regional blood flow regulation in exercising muscles.

3.2.2 $\dot{Q}$ and $\dot{V}O_2$ Kinetics

Hoffmann et al. (2016) analysed HR and $\dot{V}O_2$musc kinetics before and after ISS missions. They reported non-significantly slowed kinetics for HR and $\dot{V}O_2$pulm, but significantly slowed $\dot{V}O_2$musc kinetics 6 days after landing (Fig. 3.4). Even after 21 days, $\dot{V}O_2$musc kinetics were not completely recovered. They also compared the kinetics changes with changes in $\dot{V}O_2$peak (reported by Moore et al. 2014). Although $\dot{V}O_2$peak was already close to pre mission levels after 10 recovery days, $\dot{V}O_2$musc kinetics differed significantly from preflight values 21 days after the missions. Thus, $\dot{V}O_2$musc kinetics must be regarded as incompletely recovered up to 21 days following long-duration spaceflight. Even if the recovery countermeasures were successfully applied with regard to restoring the cardiovascular characteristics which might limit $\dot{V}O_2$peak, the training has been less successful in preventing the restoring ability of muscles to respond to transients from lower to higher energy demands.

Several ground-based studies demonstrate that assessment of these kinetics may provide valuable additional information about the system characteristics (e.g. Barstow et al. 1994; Grassi 2006). One potential reason for the small amount of data related to spaceflight might be the efforts involved in obtaining such data.

This might be related to test durations with the commonly applied step changes in WR or the complexity of interpretation of the data or combinations of these. One of the authors of the present article (Moore—also author of Moore et al. 2014) was concerned with the variation that can be associated with breath-by-breath data and preferred (at least for the first paper to define $\dot{V}O_2$peak during ISS flight) to utilise the mixing chamber method of expired gas fraction analysis. However, with the proper instrumentation, the kinetics analysis provides options for a more detailed non-invasive examination by CPET (Barstow and Molé 1987; Barstow et al. 1990; Hughson 2009).

In addition to the capacities of the physiological subsystems in O_2 transport and utilisation, the regulation processes associated with aerobic metabolism are likely subject of adaptation to weightlessness. Several studies have demonstrated changes in cardiovascular regulation under resting conditions (e.g. Cooke et al. 2000) during and after spaceflight. The typical exercise protocols that have been used for studies during spaceflight to measure $\dot{V}O_2$peak were not well suited to investigate the regulatory responses. However, the regulation of adjustments always influences the $\dot{V}O_2$ data after WR changes during exercise and may influence the final $\dot{V}O_2$peak values as well (Burnley and Jones 2007). Other interesting information can be derived from examination of $\dot{V}O_2$musc kinetics, as these kinetics are correlated with the aerobic capacity of the muscle group tested. This characteristic is independent from systemic O_2 delivery to the muscle but takes intramuscular O_2 diffusion into account (Hughson 2009).

$\dot{V}O_2$musc cannot be determined directly, and because of non-linear distortions from venous return, $\dot{V}O_2$pulm does not necessarily represent $\dot{V}O_2$musc (Barstow and Molé 1987; Eßfeld et al. 1991). This being stated, the examination of responses to WR changes can give valid information about transport and $\dot{V}O_2$musc kinetics. During CPET, typically HR and $\dot{V}O_2$pulm data are available. For a valid determination of $\dot{V}O_2$musc kinetics, several requirements must be met: (a) $\dot{Q}$ changes should represent changes of blood flow to the exercising muscles with a constant fraction being required by the non-exercising parts of the body; (b) venous volumes between muscles and lungs should be constant during the test; (c) arterial O_2 saturation should be constant; (d) and the maximal WR should be below the anaerobic threshold to approximate a linear relation between WR and $\dot{V}O_2$musc. Finally, the natural variation of breath-by-breath data requires test repetitions and/or statistical test protocols like randomised WR changes.

The results from Convertino et al. (1984) indicate slower O_2 uptake kinetics from an exercise step response after 7 days of bedrest in the supine position. However, they did not apply breath-by-breath analysis so a sophisticated kinetics analysis was not possible. In contrast, in the studies of Stegemann et al. (1997) and Hoffmann et al. (2016), beat-to-beat and breath-by-data, respectively, were analysed for HR and $\dot{V}O_2$ kinetics by applying randomised changes in WR.

Typically, rectangular changes in WR or randomised changes between two WR levels below subject's anaerobic threshold (Beaver et al. 1986) are applied during

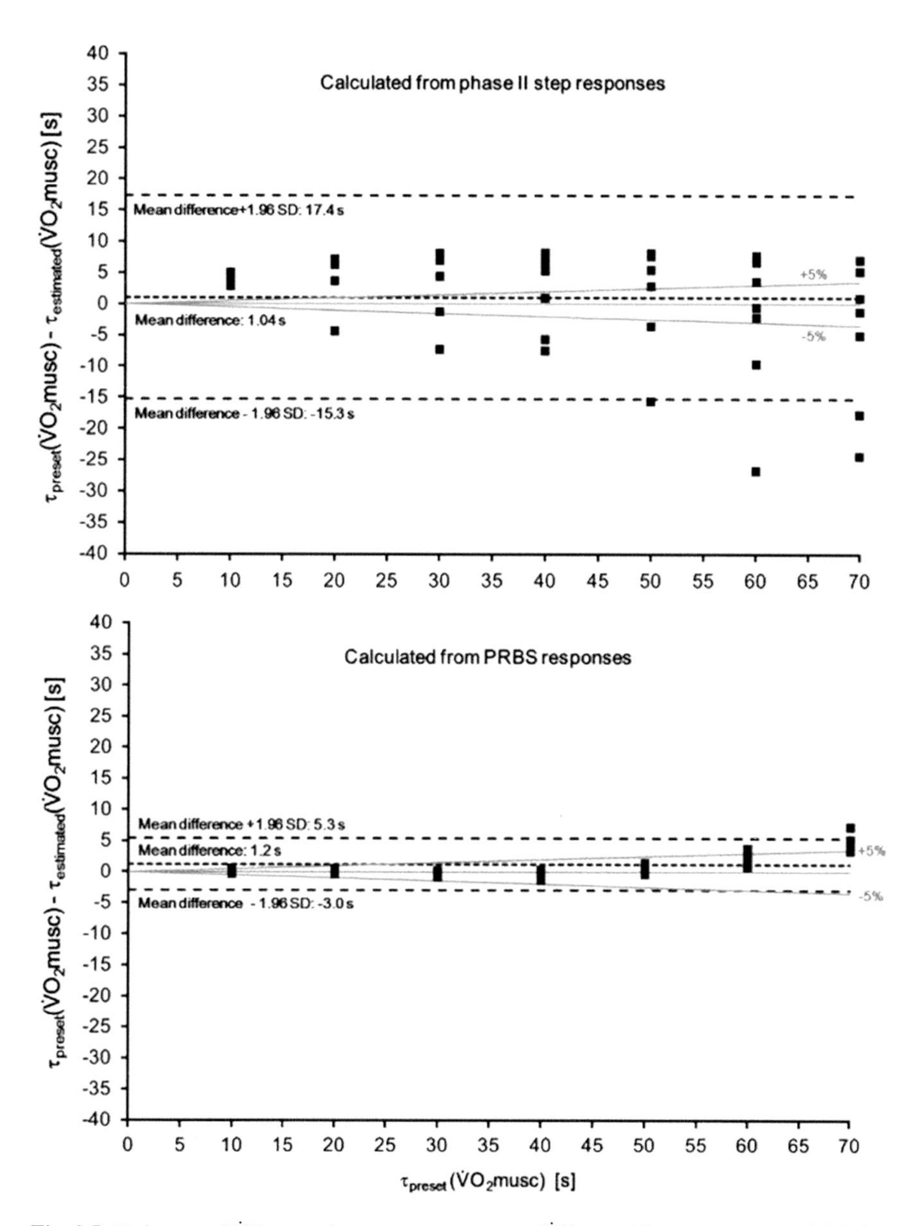

Fig. 3.5 Estimates of $\dot{V}O_2$musc time constants $\tau_{estimated}$ ($\dot{V}O_2$musc) by mono-exponential fitting of step responses of phase II (Barstow and Molé 1987, *upper panel*) and by time series analysis of PRBS responses (Hoffmann et al. 2013, *lower panel*) from simulated $\dot{V}O_2$pulm. $\dot{V}O_2$pulm responses were calculated with constant venous volume Vven = 3 L and with mono-exponential responses with τ_{preset} ($\dot{V}O_2$musc) (abscissa) and $\tau(\dot{Q})$ in the range between 10 and 70 s (see Hoffmann et al. 2013 for details)

studies of kinetics responses. From these data, the kinetics of any CPET parameter can be described, and moreover, $\dot{V}O_2$musc kinetics can be calculated either from step responses (Barstow et al. 1990) or from randomised WR changes (Hoffmann et al. 2016). While the analysis according to Barstow et al. (1990) assumed that $\dot{Q}$ does not influence the $\dot{V}O_2$musc kinetics in the second phase of step responses, Hoffmann et al. (2013) compensated for $\dot{Q}$ influences from HR data, while stroke volume was assumed to vary only in a limited range.

In Fig. 3.5, the reliability of both methods is calculated using theoretical data for both approaches. For a given time constant τ preset for $\dot{V}O_2$musc kinetics, $\dot{V}O_2$pulm data were calculated for step responses and randomised changes (pseudo randomised binary sequence, PRBS) in WR. The $\dot{V}O_2$pulm step responses were analysed by the mono-exponential fitting to the phase II responses according to the method of Barstow et al. (1990), while the $\dot{V}O_2$ and HR responses to PRBS were analysed by time series analysis as described by Hoffmann et al. (2013). These results demonstrate a higher reliability of the method described by Hoffmann et al. (2013).

In the context of spaceflight experiments, the number of repetitions of rectangular WR changes with sufficient periods of constant WR to allow a sufficient stabilisation of CPET parameters at each level would produce an unfeasible test duration (e.g. five repetitions of single-step responses would last about 1 h, without preparation of the instruments and the subject, warm up and recovery phase). Alternatively, Stegemann et al. (1985) applied the PRBS protocol in a bedrest study but analysed only the data for only $\dot{V}O_2$pulm and HR kinetics. With an improved algorithm for data analysis, $\dot{V}O_2$musc kinetics can also be assessed from these data (Hoffmann et al. 2016).

Stegemann et al. (1985) compared HR and $\dot{V}O_2$pulm kinetics before and after 7 days of bedrest. The data were evaluated using frequency analysis which complicates the interpretation. The results suggested slowed $\dot{V}O_2$pulm kinetics, but no obvious change in HR kinetics on the first day after bedrest with a recovery in $\dot{V}O_2$pulm within 5 days. During and after spaceflight, Stegemann et al. (1997) also analysed both HR and $\dot{V}O_2$pulm kinetics. They found indications of reduced venous blood volume between muscles and lungs, but they did not analyse for $\dot{V}O_2$musc kinetics.

3.3 Methodological Aspects of CPET During Spaceflight

3.3.1 Hardware

Three hardware components are essential for CPET: (a) instrumentation for respiratory gas analysis, consisting of sensors for a continuous detection of PO_2, PCO_2 and gas flow during in- and expiration, (b) a reliable and accurate method of HR monitoring and (c) an ergometer. A computer unit should be utilised to control the ergometer and to store the data from the connected components for later analysis. The computer also allows the capability for real-time monitoring (both in Space and on ground via telemetry). Other ancillary instrumentation can be added, e.g. for blood pressure measurements or for O_2 pulse oximetry.

3.3.2 Pulmonary Gas Exchange Analysis

$\dot{V}O_2$pulm is a key measurement of aerobic metabolism as O_2 has limited storage capacity in the human body and is the terminal electron acceptor in the electron transport chain of the mitochondria. $\dot{V}O_2$pulm can be measured non-invasively and in conjunction with other cardiorespiratory parameters and provides information regarding the capacities and regulations of O_2 transport and O_2 utilisation (Wasserman et al. 1999). The technique which provides the most information regarding $\dot{V}O_2$pulm measurements involves continuous detection of PO_2, PCO_2 and gas flow during in- and expiration at the mouth. From these data $\dot{V}O_2$pulm, CO_2 output ($\dot{V}CO_2$pulm), total in- and expiratory volumes, end-tidal partial O_2 and CO_2 pressures and breathing frequency can be computed breath by breath. Typically, $\dot{V}O_2$pulm and $\dot{V}CO_2$pulm are corrected for changes in lung stores of oxygen (Beaver et al. 1981). Less sophisticated instrumentations with mixing box collection of expired gases for gas analysis only allow the determination of average gas exchange for a given period, e.g. 10–30 s.

The first studies regarding aerobic metabolism and spaceflight were performed using only pre- and postflight measurements. The first in-flight measurements were conducted during the Skylab missions, but the measurements were limited to only $\dot{V}O_2$, $\dot{V}CO_2$, tidal volumes and breathing frequency (Michel et al. 1977). During the Space Shuttle era of spaceflight and continuing on the ISS, several instruments have been developed and refined to allow in-flight CPET measurements. The most current device on the ISS is the Portable Pulmonary Function System (PPFS) of the European Space Agency (ESA) (see Clemensen et al. 1994 for details). The PPFS also allows the determination of cardiac output by an inert gas rebreathing method.

The interface between a test subject and the CPET measurement device is an important consideration. The choices are typically between a mouthpiece connected to a respiratory valve, with the subject's nose occluded with a nose clip, or the use of a face mask. For the typical CPET with durations of 15 min or more, the mask is more comfortable than the combination of mouthpiece and nose clip. Salivation and the restriction to mouth breathing are typically reported as inconvenient and annoying when using the mouthpiece/nose clip combination. However, the weak points of a mask as interface between subject and gas exchange sensors are a slightly higher dead space and a higher risk for leakage (the area near the bridge of the nose is especially problematic in most masks, particularly at high expiratory volumes). To date, the mouthpiece/nose clip option has been used for spaceflight measurements.

3.3.3 Heart Rate Monitoring

The requirements for HR monitoring can vary significantly with the goal of examination. A simple bipolar lead is sufficient to determine beat-to-beat HR. For detailed cardiological examination, the detection by a 12-channel lead ECG may be necessary. A variety of instruments for HR monitoring have been qualified for spaceflight since the first spaceflight missions. Typically, ECG has been used for CPET, and during the Space Shuttle and ISS programs, technology such as that used in Polar™ heart rate monitors has been employed to use for routine exercise countermeasure monitoring (Moore et al. 1997).

3.3.4 Exercise Mode and Ergometer

The purpose of an exercise test is typically a critical factor in the selection of a testing modality. If the systemic cardiopulmonary functions are the focus of the examination, larger muscle groups must be involved. The common exercise modes on Earth are the cycle ergometer and the treadmill. For some special applications, modifications such as arm cranking are available. For intra- and interindividual comparisons, a high degree of standardisation of the exercise and the involved muscle groups are essential. The treadmill is problematic for CPET testing in Space due to the number of factors that are either difficult to control (such as an accurate and reproducible effective load applied to the exerciser though the harness system) or are unknown (such as changes in efficiency induced by exercising on an unstable surface) at this time. Although treadmills and other training devices, e.g. for strength training and upper body exercise, are used for countermeasure training, no CPET data are available for in-flight exercise and would be of interest for future study. Body position, joint angles and frequency of movements are much easier to control using a cycle ergometer. For the various studies using CPET exercise measurements in Space to date, differing set-ups have been used between investigations for subjects to interface with the cycle ergometer (Levine et al. 1996; Stegemann et al. 1997; Moore et al. 2014). Shoulder straps, direct fixation on the saddle and back support for cycle ergometry have been used.

The advantage of cycle ergometry is obviously the simplicity of subject movement which allows a high degree of standardisation. However, the differences in creation of counterforces for pedalling produce some differences between cycling on Earth versus cycling in Space. For cycling, walking and running on Earth, the counterforces are generated by the gravitational weight of the body. Movements of the upper body, the position on the ergometer and the resulting working angles may influence the efficiency (Too 1990) and, therefore, the CPET results. One disadvantage of cycle ergometry is an issue of exercise specificity. Ambulation is an important component of what will be required for a Space traveller following flight. Ambulation utilises some differing muscles than cycling or similar muscles but in an entirely different motor pattern. From this perspective, treadmill exercise has an advantage compared to cycling.

For any ergometer used in Space, the transmission of vibrations to the structures of the covering module has to be minimised. The Cycle Ergometer with Vibration Isolation and Stabilization (CEVIS) and the Treadmill with Vibration Isolation System (TVIS), designed for the ISS, are typical examples of Space optimised devices for testing and training. For treadmills two types can be distinguished: motor- versus non-motor-driven devices (Lakomy 1987). The Russian treadmill (BD-2) on-board the ISS allows both options.

Both exercise modes, cycling and treadmill, require large muscle groups and can give valid information about systemic limitations of O_2 transport. For testing of smaller muscle groups, special types of ergometers were designed and utilised to study the cardiovascular responses to exercise (McCreary et al. 1996; Eßfeld et al. 1993).

3.3.5 Exercise Test Protocols

The selection of an adequate WR protocol is related to the aim of the investigation (see also Sect. 3.2). To some degree, the selection of the protocol is a compromise of optimal test procedure, influenced by physiological and psychological limitations along with the feasibility of conducting the protocol (Fig. 3.6). The aims of the

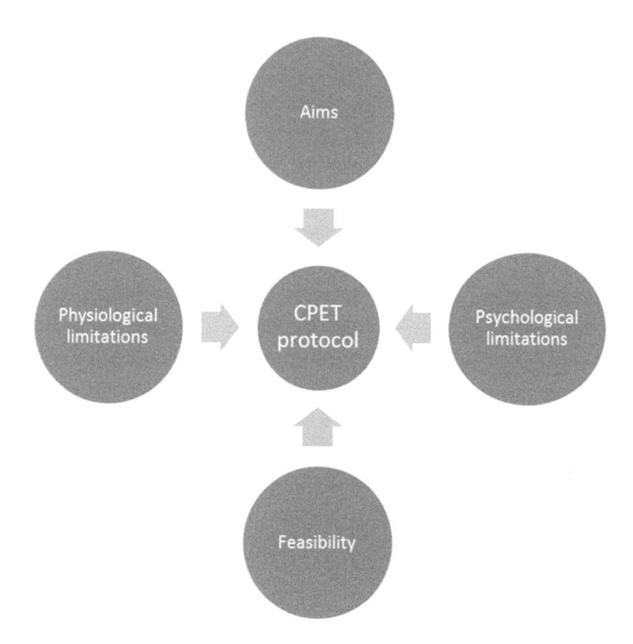

Fig. 3.6 Criteria for the selection of an exercise protocol

CPET can be the assessment of systemic respiratory, cardiovascular and muscular capacities as well as the physiological regulation of these factors (Wasserman et al. 1999). Estimates of gross efficiency may also be derived from a CPET (Gaesser and Brooks 1975). Examples of physiological limitations are muscular strength of the subject and the duration of a CPET, which may cause fatigue prior to the attainment of a physiological maximum. Although a minimum of subject's collaboration and desire is obligatory, some testing situations require a significant level of motivation. This is especially true in cases during which the WR are close to the maximum of subject's physical performance.

Safety of the tests is of great importance. For CPET, especially for those conducted in Space, the probability of requiring direct medical support must be minimised. Other factors to consider are the duration of time required to set and calibrate equipment to support the testing session and the time for recovery.

A wide variety of exercise tests exists. Sinusoidal changes, square waves, randomised changes, stepwise increases, ramp and constant WRs are some examples of work rate perturbations that can be performed during a test (Casaburi et al. 1977; Linnarsson 1974; Hughson and Inman 1986; Eßfeld et al. 1987). Also combinations of these work rate applications within a test protocol are possible. For any increasing WR, the amplitudes and the resulting forces, which are related to the ergometer type, must be carefully considered. For cycle ergometry, pedal speed should be constant if the WR is considered to be a measure of induced metabolism, since efficiency varies with pedal speed (Gaesser and Brooks 1975). The given pedal speed for most CPET protocols should be within the range between 70 and 90 rpm to avoid high muscle forces and insufficient muscle perfusions at lower speeds and coordinative problems, especially for subjects with no cycling experience.

Another criterion for the protocol selection and the selection of proper instrumentation is the nature of the subsequent data analysis envisioned for the CPET. For any physiological data, i.e. breath-by-breath data, typical variations must be accounted for (Lamarra et al. 1987). For some purposes, such as steady-state estimates, it is possible to reduce variation by the use of a moving average. Integration of the response data for a given time interval within a constant WR is another method whereby variation can be reduced. This approach is typical for the analysis of incremental exercise tests to determine the relationship of the physiological variables, e.g. $\dot{V}O_2$pulm versus $\dot{V}CO_2$pulm or WR versus $\dot{V}O_2$pulm. However, for the assessment of cardiorespiratory kinetics after WR changes, this type of analysis is equivalent to low-pass filtering and, therefore, not appropriate. Alternatively, to minimise natural fluctuations of CPET data in the transient phases, repetitions of the test can be performed with subsequent averaging of the data. This is not feasible in many circumstances, including during spaceflight experiments where time is a premium commodity. A variation of this approach is to require temporary steady states at least for the starting WR, which increases the test duration. Another alternative is given by time series analysis which requires specific WR protocols, such as sinusoidal or randomised WR changes (Casaburi et al. 1977; Eßfeld et al. 1987).

Frequently applied WR protocols are stepwise or nearly continuous increases of WR up to submaximal levels or criteria-defined limits (e.g. a "termination heart rate") with increments of 15–20 W every 2 min (Albouaini et al. 2007). These protocols may be appropriate for some clinical tests for use with severely deconditioned subjects; however, steeper increases are recommended for most apparently healthy individuals to avoid termination by local muscular fatigue caused by high proportions of anaerobic metabolism late in the test. This would result in the subject being unable to achieve sufficient cardiorespiratory responses. The spiroergometric data collected during a protocol with the proper work progression can yield valuable information about the characteristics of the muscle groups tested such as the anaerobic threshold related to WR (see Albouaini et al. 2007 for an overview). Moreover, regression analysis of WR with properly collected cardiorespiratory parameters allows a prescription of optimised training intensities. Cardiopulmonary regulation is typically not the focus of these incremental tests.

3.4 Conclusions

Future exercise monitoring during and following spaceflight should involve measurement of oxygen uptake kinetics as well as maximal systemic aerobic capacity. Since exercise tests for kinetics analysis are not influenced by factors such as the influence of mental fatigue or subject enthusiasm, which can influence maximal testing results, kinetics testing might be easier to integrate during Space missions. Nevertheless, for a properly combined examination of both aerobic exercise capacity and the regulatory mechanisms contributing to aerobic performance, a pseudo randomised binary sequence exercise protocol should be combined with the graded exercise test. The PRBS would serve as a useful data collection period and would easily be incorporated as a "warm up" to the subsequent more traditional work rate increases of a CPET.

References

Albouaini K, Egred M, Alahmar A, Wright DJ (2007) Cardiopulmonary exercise testing and its application. Heart 93:1285–1292

Barstow TJ, Molé PA (1987) Simulation of pulmonary O_2 uptake during exercise. J Appl Physiol 63:2253–2261

Barstow TJ, Scheuermann BW (2005) V'O_2 kinetics: effects of maturation and ageing. In: Jones AM, Poole DC (eds) Oxygen uptake kinetics in sport, exercise and medicine. Routledge, Abingdon

Barstow TJ, Lamarra N, Whipp BJ (1990) Modulation of muscle and pulmonary O_2 uptakes by circulatory dynamics during exercise. J Appl Physiol 68:979–989

Barstow TJ, Buchthal S, Zanconato S, Cooper DM (1994) Muscle energetics and pulmonary oxygen uptake kinetics during moderate exercise. J Appl Physiol 77:1742–1749
Beaver WL, Wasserman K, Whipp BJ (1986) A new method for detecting anaerobic threshold by gas exchange. J Appl Physiol 60:2020–2027
Beaver WL, Lamarra N, Wasserman K (1981) Breath-by-breath measurement of true alveolar gas exchange. J Appl Physiol Respir Environ Exerc Physiol 51:1662–1675
Burnley M, Jones AM (2007) Oxygen uptake kinetics as a determinant of sports performance. Eur J Sport Sci 7:63–79
Casaburi R, Whipp BJ, Wasserman K, Beaver WL, Koyal SN (1977) Ventilatory and gas exchange dynamics in response to sinusoidal work. J Appl Physiol Repir Environ Exerc Physiol 42:300–311
Chi MM, Hintz CS, Coyle EF, Martin WH, Ivy JL, Nemeth PM, Holloszy JO, Lowry OH (1983) Effects of detraining on enzymes of energy metabolism in individual human muscle fibers. Am J Physiol-Cell Physiol 244:C276–C287
Clemensen P, Christensen P, Norsk P, Gronlund J (1994) A modified photo- and magnetoacoustic multigas analyzer applied in gas exchange measurements. J Appl Physiol 76:2832–2839
Convertino VA (1997) Cardiovascular consequences of bed rest: effect on maximal oxygen uptake. Med Sci Sports Exerc 29:191–196
Convertino VA, Goldwater DJ, Sandler H (1984) VO_2 kinetics of constant-load exercise following bed-rest-induced deconditioning. J Appl Physiol Respir Environ Exerc Physiol 57:1545–1550
Convertino V, Hung J, Goldwater D, DeBusk RF (1982) Cardiovascular responses to exercise in middle-aged men after 10 days of bedrest. Circulation 65:134–140
Cooke WH, Ames JE IV, Crossman AA, Cox JF, Kuusela TA, Tahvanainen KUO, Moon LB, Drescher J, Baisch FJ, Mano T, Levine BD, Blomqvist CG, Eckberg DL (2000) Nine months in space: effects on human autonomic cardiovascular regulation. J Appl Physiol 89:1039–1045
Eßfeld D, Hoffmann U, Stegemann J (1987) $\dot{V}O_2$ kinetics in subjects differing in aerobic capacity: investigation by spectral analysis. Eur J Appl Physiol 56:508–515
Eßfeld D, Hoffmann U, Stegemann J (1991) A model for studying the distortion of muscle oxygen uptake patterns by circulation parameters. Eur J Appl Physiol 62:83–90
Eßfeld D, Baum K, Hoffmann U, Stegemann J (1993) Effects of microgravity on interstitial muscle receptors affecting heart rate and blood pressure during static exercise. Clin Investig 71:704–709
Ferretti G, Antonutto G, Denis C, Hoppeler H, Minetti AE, Narici MV, Desplanches D (1997) The interplay of central and peripheral factors in limiting maximal O_2 consumption in man after prolonged bed rest. J Physiol 501:677–686
Fortney SM, Hyatt KH, Davis JE, Vogel JM (1991) Changes in body fluid compartments during a 28-day bed rest. Aviat Space Environ Med 62:97–104
Gaesser GA, Brooks GA (1975) Muscular efficiency during steady-state exercise: effects of speed and work rate. J Appl Physiol 89:1132–1138
Grassi B (2006) Oxygen uptake kinetics: why are they so slow? And what do they tell us? J Physiol Pharmacol 57(Suppl 10):53–65
Greenleaf JE (1984) Physiological responses to prolonged bed rest and fluid immersion in humans. J Appl Physiol 57:619–633
Hoffmann U, Drescher U, Benson AP, Rossiter HB, Essfeld D (2013) Skeletal muscle $\dot{V}O_2$ kinetics from cardiopulmonary measurements: assessing distortions through O_2 transport by means of stochastic work-rate signals and circulatory modelling. Eur J Appl Physiol 113:1745–1754
Hoffmann U, Moore AD, Koschate J, Drescher U (2016) V´O and HR kinetics before and after international space station missions. Eur J Appl Physiol 116:503–511
Hughson RL (2009) Oxygen uptake kinetics: historical perspective and future directions. Appl Physiol Nutr Metabol 34:840–850
Hughson RL, Inman MD (1986) Oxygen uptake kinetics from ramp work tests: variability of single test values. J Appl Physiol 61:373–376

Klausen K, Andersen LB, Pelle I (1981) Adaptive changes in work capacity, skeletal muscle capillarization and enzyme levels during training and detraining. Acta Physiol Scand 113:9–16
Lakomy HKA (1987) The use of a non-motorized treadmill for analysing sprint performance. Ergonomics 30:627–637
Lamarra N, Whipp BJ, Ward SA, Wasserman K (1987) Effect of interbreath fluctuations on characterizing exercise gas exchange kinetics. J Appl Physiol 62:2003–2012
Levine BD, Lane LD, Watenpaugh D, Gaffney F, Buckey JC, Blomqvist CG (1996) Maximal exercise performance after adaptation to microgravity. J Appl Physiol 81:686–694
Leyk D, Eßfeld D, Hoffmann U, Wunderlich HG, Baum K, Stegemann J (1994) Postural effect on cardiac output, oxygen uptake and lactate during cycle exercise of varying intensity. Eur J Appl Physiol 68:30–35
Linnarsson D (1974) Dynamics of pulmonary gas exchange and heart rate changes at start and end of exercise. Acta Physiol Scand Suppl 415:1–68
McCreary CR, Chilibeck PD, Marsh GD, Paterson D, Cunningham DA, Thompson RT (1996) Kinetics of pulmonary oxygen uptake and muscle phosphates during moderate-intensity calf exercise. J Appl Physiol 81:1331–1338
Michel EL, Rummel JA, Sawin CF, Buderer MC, Lem JD (1977) Results of Skylab experiment M171 – metabolic activity. In: Johnson RS, Dietlein LF (eds) Biomedical results from Skylab, NASASP-377. National Aeronautics and Space Administration, Washington, DC, pp 372–387
Midgley AW, McNaughton LR, Polman R, Marchant D (2007) Criteria for determination of maximal oxygen uptake. Sports Med 37:1019–1028
Moore AD, Lee SM, Greenisen MC, Bishop P (1997) Validity of a heart rate monitor during work in the laboratory and on the space shuttle. Am Ind Hyg Assoc J 58:299–301
Moore AD, Lee SMC, Stenger MB, Platts SH (2010) Cardiovascular exercise in the U.S. space program: past, present, and future. Acta Astonaut 66:974–988
Moore AD, Downs ME, Lee SMC, Feiveson AH, Knudsen P, Ploutz-Snyder L (2014) Peak exercise oxygen uptake during and following long-duration spaceflight. J Appl Physiol 117:231–238
Mujika I, Padilla S (2001) Cardiorespiratory and metabolic characteristics of detraining in humans. Med Sci Sports Exerc 33:413–421
Norsk P, Stadeager C, Johansen LB, Warberg J, Bie P, Foldager N, Christensen NJ (1993) Volume-homeostatic mechanisms in humans during a 12-h posture change. J Appl Physiol 75:349–356
Pavy-Le Traon A, Heer M, Narici MV, Rittweger J, Vernikos J (2007) From space to earth: advances in human physiology from 20 years of bed rest studies (1986–2006). Eur J Appl Physiol 101:143–194
Perhonen MA, Franco F, Lane LD, Buckey JC, Blomqvist CG, Zerwekh JE, Peshocl RM, Weatherall PT, Levine BD (2001) Cardiac atrophy after bed rest and spaceflight. J Appl Physiol 91:645–653
Stegemann J, Essfeld D, Hoffmann U (1985) Effects of a 7 day head-down Tilt ($-6°$) on the dynamics of oxygen uptake and heart rate adjustment in upright exercise. Aviat Space Environ Med 56:410–414
Stegemann J, Hoffmann U, Erdmann R, Eßfeld D (1997) Exercise capacity during and after spaceflight. Aviat Space Environ Med 68:812–817
Too D (1990) Biomechanics of cycling and factors affecting performance. Sports Med 10:286–302
Trappe T, Trappe S, Lee G, Widrick J, Fitts R, Costill D (2005) Cardiorespiratory responses to physical work during and following 17 days of bed rest and spaceflight. J Appl Physiol 100:951–957
Tuday EC, Meck JV, Nyhan D, Shoukas AA, Berkowitz DE (2007) Microgravity-induced changes in aortic stiffness and their role in orthostatic intolerance. J Appl Physiol 102:853–858
Wagner PD (1995) Muscle O_2 transport and O_2 dependent control of metabolism. Med Sci Sports Exerc 27:47–53

Wasserman K, Hansen JE, Sue DY, Whipp BJ, Casaburi R (1999) Principles of exercise testing and interpretation: including pathophysiology and clinical applications, 3rd edn. Lippincott Williams & Wilkins, Philadelphia

Related but not Cited

Bassett DR Jr, Howley ET (2000) Limiting factors for maximum oxygen uptake and determinants. Med Sci Sports Exer 32:70–84

Capelli C, Antonutto G, Kenfack MA, Cautero M, Lador F, Moia C (2006) Factors determining the time course of V'O_2max decay during bedrest: implications for V'O_2max limitation. Eur J Appl Physiol 98:152–160

Cochrane JE, Hughson RL (1992) Computer simulation of O_2 transport and utilization mechanisms at the onset of exercise. J Appl Physiol 73:2382–2388

Convertino VA, Keil LC, Bernauer EM, Greenleaf JE (1981) Plasma volume, osmolality, vasopressin, and renin activity during graded exercise in man. J Appl Physiol 50:123–128

DeLorey SD, Kowalchuk JM, Paterson DH (2004) Effects of prior heavy-intensity exercise on pulmonary O_2 uptake and muscle deoxygenation kinetics in young and older adult humans. J Appl Physiol 97:998–1005

DeLorey DS, Paterson DH, Kowalchuk JM (2007) Effects of ageing on muscle O_2 utilization and muscle oxygenation during the transition to moderate-intensity exercise. Appl Physiol Nutr Metab 32:1251–1262

Di Prampero PE, Davies DTM, Crretelli P, Margaria R (1970) An analysis of O_2 debt contracted in submaximal exercise. J Appl Physiol 29:547–551

Drescher U, Koschate J, Hoffmann U (2015) Oxygen uptake and heart rate kinetics during dynamic upper and lower body exercise: an investigation by time-series analysis. Eur J Appl Physiol 115:1665–1672. doi:10.1007/s00421-015-3146-4

Hughson RL, Sherrill DL, Swanson GD (1988) Kinetics of VO_2 with impulse and step exercise in humans. J Appl Physiol 64:451–459

Inman MD, Hughson RL, Weisiger KH, Swanson GD (1987) Estimate of mean tissue O_2 consumption at onset of exercise in males. J Appl Physiol 63:1578–1585

Lee SMC, Williams WJ, Schneider SM (2002) Role of skin blood flow and sweating rate in exercise thermoregulation after bed rest. J Appl Physiol 92:2026–2034

Lee SM, Moore AD, Everett ME, Stenger MB, Platts SH (2010) Aerobic exercise deconditioning and countermeasures during bed rest. Aviat Space Environ Med 81:52–63

Moore AD Jr, Lee SM, Charles JB, Greenisen MC, Schneider SM (2001) Maximal exercise as a countermeasure to orthostatic intolerance after spaceflight. Med Sci Sports Exerc 33:75–80

Petrini MF, Peterson BT, Hyde RW (1978) Lung tissue volume and blood flow by rebreathing theory. J Appl Physiol 44:795–802

Richardson RS, Noyszewski EA, Kendrick KF, Leigh JS, Wagner PD (1995) Myoglobin O_2 desaturation during exercise – evidence of limited O_2 transport. J Clin Invest 96:1916–1926

Rossiter HB, Ward SA, Doyle VL, Howe FA, Griffiths JR, Whipp BJ (1999) Inferences from pulmonary O_2 uptake with respect to intramuscular [phosphocreatine] kinetics during moderate exercise in humans. J Physiol 518:921–932

Trappe S, Costill D, Gallagher P, Creer A, Peters JR, Evans H, Riley DA, Fitts RH (2009) Exercise in space: human skeletal muscle after 6 months aboard the international space station. J App Physiol 106:1159–1168

Tschakovsky ME, Hughson RL (1999) Interaction of factors determining oxygen uptake at the onset of exercise. J Appl Physiol 86:1101–1113

Whipp BJ, Ward SA, Lamarra N, Davis JA, Wasserman K (1982) Parameters of ventilatory and gas exchange dynamics during exercise. J Appl Physiol 52:1506–1513

Chapter 4
Enhancing Mental Health: Effects of Exercise on Social Well-Being and Social Ill-Being

Fabian Pels and Jens Kleinert

Abstract The impact of the social environment on health seems to be particularly important under difficult environmental conditions (e.g. Space flight). This chapter presents three studies evaluating the moderating function of social relationships and competitive settings on the impact of the social environment in the context of exercise on social well-being and ill-being. In general, these studies indicate that activities aiming at increasing social well-being and decreasing social ill-being can be cooperative or competitive in nature. However, competitive exercise activities should only be conducted if the relationship between the involved individuals is already positive before the competition takes place. Future studies need to test whether the existing exercise facilities in Space flight (e.g. ergometers, steppers) can actually be used for an increase in social well-being.

Keywords Social well-being • Loneliness • Aggression • Interdependence • Need satisfaction

4.1 Introduction

By 1946, the WHO was taking into account the tripartite of physical, mental and social well-being when defining health. Accordingly, social well-being is now considered as an important part of the overall well-being of a person. However, whereas mental well-being has extremely often been considered as an issue in research, the topic of social well-being has seldom been a focus point. One reason for this lack of research into social well-being might be that the definition and, thus, the operationalization of social well-being is often imprecise; a further reason might be the unclear distinction between it and mental well-being, as both social and mental well-being are subjective experiences of individuals.

Despite this lack of research and the frequent weakness of a theoretical conceptualization of social well-being, there exists apparently a strong connection between social well-being and mental health. According to Keyes (1998, p. 122), this

F. Pels • J. Kleinert (✉)
German Sport University Cologne, Am Sportpark Müngersdorf 6, 50933 Köln, Germany
e-mail: f.pels@dshs-koeln.de; Kleinert@dshs-koeln.de

S. Schneider (ed.), *Exercise in Space*, SpringerBriefs in Space Life Sciences,
DOI 10.1007/978-3-319-29571-8_4

connection makes it necessary to investigate adults' social well-being in order to understand optimal functioning and mental health. In such investigations, the social environment obviously plays an important role, while significant others like friends, colleagues, teachers or coaches are able to provoke an enhancement or a decrease of well-being and mental health. Thus, the social environment and social interaction are both one of the greatest contributors and one of the greatest threats to mental health.

4.1.1 Social Environment and Mental Health in Space

The impact of social circumstances on mental health seems to be particularly important under difficult or challenging social conditions. Work groups in Space, in particular, are a good example of such challenging group conditions (see Kanas 2005; Kanas et al. 2009 for an overview). For instance, intercultural conditions can be difficult because group members in Space labs typically come from different countries. Moreover, due to the long-term isolated group living, the quality of interpersonal bonding seems to be very important. Finally, individuals in Space labs strongly depend on each other for the success of their working tasks and goal attainment, which require teamwork, cooperation and communication.

One could say that work groups in Space are, essentially, social-psychological group experiments in themselves, while they also benefit from social-psychological research into the dynamics of these specific living conditions. Considering the latter, research on treatments that enhance well-being and functioning of both individuals and groups may be transferable to living and working in Space. The efficacy of such a transfer depends on whether studies or experiments on earth consider special conditions or at least typical situations or situational determinants similar to those found in Space. Besides this Earth-Space transfer, it is important to note that such research and existing knowledge of social-psychological mechanisms may also help to find ways to solve general interpersonal problems in society (e.g. aggression, loneliness).

With the relationship between social environment, social well-being and mental health in mind (particularly but not exclusively in Space), the aim of this chapter is to discuss the potentially positive role of exercise within this relationship. Towards this aim, the chapter is structured into a theoretical and an empirical part. The theoretical part (1) defines the concept of social well-being, explains effects of exercise on (2) social well-being and (3) ill-being and (4) discusses moderators of the relation between exercise and social well-being and ill-being, respectively.

The empirical part of the chapter presents three studies with different approaches to the relationship between social environment and social well-being, all based on exercise-related treatments. The first study is on the enhancement of social well-being in different exercise tasks; the second focuses on the reduction of loneliness (as a form of social ill-being) via exercise; and the third examines the reduction of aggression (as a typical social emotion) via different forms of exercise. All studies are compared in a general discussion at the end of the chapter.

4.1.2 Definition of Social Well-Being

Like mental well-being or physical well-being, social well-being is essentially for an individual's affect. Accordingly, social well-being can be defined as a subjective experience of a specific mental state, which is characterized by a positive or negative valence (i.e. a positive or negative mood state). However, in contrast to mental well-being or physical well-being, social well-being is determined by social aspects. Thus, Keyes defines social well-being as "the appraisal of one's circumstances and functioning in society" (Keyes 1998, p. 122). Taking into account the definition of social well-being as a mood state, these kinds of appraisals are not the object of social well-being directly, but they are the cause of an individual's subsequent mental state which appears as social well-being.

Regarding the target of social-oriented appraisals, Larson (1993) distinguishes between social adjustment and social support. Social adjustment comprises three aspects, namely, the satisfaction with relationships, the performance of oneself in all kinds of social roles and all forms of adjustment to (social) environments. Social support can be divided into the number of contacts and the appraisal of these contacts. In general, the structure from Larson is based on an individual's appraisal of his or her social interactions (e.g. role taking, being socially supported) and might be a function of the individual's desirable level of quality of these interactions.

Whereas Larson views social appraisals in association with the evaluation of social interactions, Keyes' (1998) perspective of appraisal issues has a more global approach (i.e. the appraisal of society). The author distinguishes between five dimensions, of which three characterize the appraisal of society itself and two the appraisal of one's belonging to society. Regarding the former, a person (1) evaluates the potential of society including its structure or trajectory (social actualization), (2) accepts society as a generalized category (social acceptance) and (3) perceives society as coherent and senseful (social coherence). Further to this, and considering the latter two dimensions, Keyes distinguishes between an individual's appraisal of his or her (4) social integration (e.g. relationships) and (5) social contribution (i.e. one's social value as a vital member of society).

Both conceptualizations have overlapping components and relationships. Overlaps exist in regard to social relationships [which are relevant in social adjustment (Larson) and social integration (Keyes)] and an individual's impact on society [which is a matter of social performance (Larson) and social contribution (Keyes)]. The most important relationship of both structures is that Keyes' view of social actualization, social acceptance and social coherence is a function of social interaction. In other words, an individual's appraisal of society in general is a function of his or her social living and experience of society (i.e. other people, organizations, social networks) while interacting in society (Simon 2004). *Thus, in the present chapter, we take social interaction as the basic source of social well-being, whereas appraisal of social interactions may moderate this relationship.*

The concepts of Keyes and Larson include an individual's appraisal and evaluation as key processes in the development of social well-being. Accordingly, the development of social well-being is mainly cognitively driven. In such a way, social satisfaction would be a rational process due to the comparison of personal goals, values or requirements on the one hand and the perception and appraisal of the personal situation on the other hand.

A different view is described by Deci and Ryan (2008). The authors explain the development of well-being as a consequence of need satisfaction. A person whose basic and innate psychological needs are met feels well. This coincidence of a positive, balanced feeling and the satisfaction of basic needs have already been described by Lewin (1943) in his broader concept of field theory. Given this theoretical consideration, social well-being can be described as the satisfaction of the basic psychological needs by or in a certain social context (Kleinert 2012, 2014). According to Deci and Ryan (2000), these basic psychological needs are the needs for autonomy, competence and relatedness. *Thus, social well-being is a function of the degree to which social interactions in a certain social context satisfy basic psychological needs.*

However, even if the satisfaction of basic psychological needs is the most important source of social well-being, rational processes of appraisal and evaluation explain a part of social well-being as well (Keyes 1998). Both field theory (Lewin 1943) and organismic integration theory (Deci et al. 1994) help to explain such results. The individual's evaluation of his own behaviour as senseful, meaningful and congruent to one's own goals and values leads to a feeling of being self-determined, balanced and internally consistent. These feelings are strongly associated with a positive mood state and well-being. As far as this perceived congruency and consistency include the interaction with the social environment, this is a matter of social well-being. In summary, besides the socially driven satisfaction of basic psychological needs, the appraisal of social interaction as being congruent with personal goals, values and important self-reflections is a second substantial source of social well-being.

4.1.3 Exercise and Social Well-Being

Exercise and physical activity have an overall positive effect on well-being and mental health (for an overview, see Biddle and Mutrie 2008; Paluska and Schwenk 2000). This is particularly well documented in cases of effects of moderate intensive exercise on positive mood state. The results of studies on vigorous activities and specific forms of sport or exercise (e.g. strength training; Biddle 1995; Glenister 1996) are less consistent. Depending on intensity and form of activities, the positive effect of exercise or physical activity is discussed as a result of different psychological and physiological mechanisms (Paluska and Schwenk 2000; Schlicht 1995). As this chapter focuses on social well-being, the present section discusses the role of these mechanisms in regard to the social interaction during exercise or sport activities. Consequently, pure neurological or physiological hypotheses (e.g. the monoamine hypothesis, the endorphin hypothesis, the thermogenic model or the visceral afferent feedback hypothesis; Paluska and Schwenk 2000) are not discussed as they do not consider specific social circumstances.

Besides neurophysiological hypotheses, Paluska and Schwenk (2000) give an overview on different psychosocial approaches in explaining exercise effects on well-being and mental health. The authors distinguish between the distraction hypothesis, the efficacy model, the mastery model and the social interaction hypothesis (see also Biddle and Mutrie 2008; Schlicht 1995). Each of these explanation models includes a specific application in social interaction settings:

(a) According to the distraction hypotheses, exercise leads to a diversion from unpleasant stimuli (e.g. pain or life problems). The framework of this model is the competition of cues approach (Pennebaker 1982), in which external stimuli are more important than internal stimuli. This predominance of external stimuli is especially relevant in social interactions in which an individual has to concentrate on his or her social partner while interacting with him or her during an exercise task or a shared activity. However, this process explains just the effect of social interaction on mental well-being in general but not a specific benefit of exercise on social well-being in particular.
(b) In self-efficacy theory, positive mood states are associated with feelings of competence in anticipating or regulating an action (Paluska and Schwenk 2000). Transferred to social settings, this positive feeling (i.e. social well-being) is caused by perceived competencies in interacting with a partner or within a group. The person perceives himself or herself as self-efficient in regard to social skills or social competencies which are necessary to interact (e.g. communication, self-presentation). This view is close to Larson's (1993) construction of social well-being as a consequence of positive role experience: a positive perception of role taking in a specific context is a matter of competence in filling out this role (e.g. acting role congruent). Thus, exercise-related social well-being as a consequence of self-efficacy beliefs depends on the given roles and one's competence to fulfil such roles by using social skills in an appropriate and role-adequate manner. In regard to exercise settings, both aspects, the different social functions or social roles within an exercise group and the social skills given by the group members, are therefore objects of the enhancement of social well-being of group individuals.
(c) In contrast to the efficacy model, the mastery hypothesis of well-being focuses more on the outcome of an action rather than on the efficacy beliefs about skills or competencies needed for the action itself. That is, the mastery hypothesis assumes that people feel good because they experience themselves as responsible for action outcomes or results (i.e. control belief). Within this concept, the object of mastering can be the social environment (Keyes

et al. 2002) or individual tasks (Ryan and Deci 2001). As far as social settings in sport are concerned, social well-being consequently appears if the individual believes that he or she is able to reach important goals or objectives together with the group. Such important goals can both be individually important and important for the group as a whole.

It should be mentioned that the mastery hypothesis of well-being is sometimes linked to the idea of mastery transfer (Schlicht 1995). Mastery transfer means that a person transfers the beliefs of control from one context or setting to another or even to a global level. Specifically, if a person experiences control and mastering in the context of sport, he or she evaluates other contexts, or his or her life as a whole, as controllable as well. This transferred or overall feeling of control is accompanied by positive well-being. In regard to social well-being, this would mean that experiences of mastery in an exercise group would lead to feelings of mastery or coping in social settings in general, which concurrently enhances social well-being in general.

(d) According to (Paluska and Schwenk 2000, p. 176) "social interaction hypothesis postulates that the social relationships and mutual support which exercisers provide each other with account for a substantial portion of the effects of exercise on mental health". This quote shows that a positive experience from the social interaction itself can lead to enhanced well-being, regardless of the evaluation of their own role or benefits which are linked to this interaction. Since these interactions are inherently social, the well-being which is connected to it is experienced as "social well-being".

Moreover, social interactions can lead to a strengthened social identity which, in turn, may enhance social well-being (Turner et al. 1987). A necessary condition of this process is the individual's perception of similarities in comparison to the interacting other. Also, in sport and exercise groups, such connection between well-being and the perception of similarities could be shown, with group identification mediating this relationship in this mechanism (Zepp and Kleinert 2015).

4.1.4 Exercise and Social Ill-Being

Ill-being can be characterized as the opposite extreme position to mental well-being (Ryff et al. 2006). Thus, according to the aforementioned definition of social well-being, social ill-being can be characterized as a function of the degree to which social interactions in a certain social context lead to a disregard or even frustration of basic psychological needs. Regarding exercise settings, frequent conflicts and socially induced anger or aggressiveness are typical examples for sport or exercise situations, in which psychological needs (i.e. autonomy, competence, relatedness) are frustrated. If need frustration takes place over a longer period or even in many life contexts, it can be expected that personal growth and mental health are thwarted (Vansteenkiste and Ryan 2013).

Up to now, the relationship between exercise or sport and ill-being is rarely considered in research, and findings in this area appear inconsistently (see, for instance, Williams and Gill 2000). It can be assumed that the underlying processes are similar to those mechanisms that explain the development of well-being (as described before); however, since ill-being is not the absence of well-being, such an assumption has to be proved in future research.

Moreover, it is not clear which specific psychological states are associated with feelings of social ill-being. However, language includes expressions for such social ill-being (e.g. "rejected", "excluded", "disregarded" or "misunderstood"). Finally, such situations are allied to other social-psychological constructs (i.e. aggression, conflict, loneliness), which thus have to be considered in future research on the relationship between exercise and social ill-being.

4.1.5 Moderators of the Relationship Between Exercise and Well-Being or Ill-Being

It is obvious that exercise does not lead to social well-being or social ill-being per se but is apparently moderated by different variables. This range of assumed variables can be categorized into (1) exercise-related variables, (2) person-related variables and (3) group-related variables.

Exercise-Related Variables Regarding exercise, movement type, exercise intensity and exercise-related interactions should be considered as probable moderators of the exercise-well-being link. Whereas movement type and exercise intensity seem to be important moderator variables for well-being in general (e.g. Biddle and Mutrie 2008), exercise-related interactions contain social characteristics which could be specifically relevant for an enhancement of social well-being. Accordingly, it might depend on the type of exercise task, in which way and to what extent it is required that group members work together and cooperate (instead of competing against each other). Such characteristics define the extent of interdependence within a given exercise task.

Person-Related Variables Whether a person is willing and able to cooperate with other group members is a function of trait and state variables of that person. As far as traits are concerned, a person with missing social skills or an introverted personality is assumed to gain less benefit from social-oriented exercises compared to an individual who is open-minded or has good social competences (Barrick et al. 1998). Regarding psychological states, negative affective states (e.g. tiredness or depression) can be expected as negative moderators of an exercise-social-well-being link since these negative mood states reduce the ability to interact.

Group-Related Variables These variables describe the given structure of a group. As mentioned before, existing roles are important moderators of the enhancement of social well-being in exercise settings. Such roles depend on formal and informal functions within a group. Leadership research in sport makes evident that both formal leaders (e.g. coach, instructor, team captain) and informal leaders (Fransen et al. 2014) are responsible for the communication and interaction within a group and, thus, have an influence on the development of social well-being and ill-being.

Moreover, the social climate within a group is an important group-related variable. In sport and exercise research, particularly, the motivational climate has been widely investigated: Motivational climate can be described as the social situation created by significant others in terms of the achievement goals being emphasized (Duda and Balaguer 2007). These achievement goals are typically subdivided into task goals (i.e. performing well during a certain task) and ego goals (i.e. being better than others). Whether the significant others (e.g. an instructor or other group members) create a task-involving or an ego-involving motivational climate during a sport or exercise, activity influences the social well-being. Both coach-created and peer-created task-involving climates positively predict the satisfaction of the needs for autonomy, competence and relatedness (e.g. Ntoumanis et al. 2007; Reinboth and Duda 2006). Conversely, an ego-oriented motivational climate thwarts need satisfaction.

Given this theoretical concept and these empirical findings, the following empirical part of the chapter presents three studies with different approaches to the relationship of social environment in the context of exercise and social well-being. In these three studies, two forms of moderators are considered: Two studies (study 1, study 2) look at mechanisms of *social relationship* as a moderator of the link between exercise and social well-being. Furthermore, two studies (study 1, study 3) deal with effects of *competitive settings* (in regard to cooperative settings or individualized settings).

4.2 Study 1: Enhancement of Social Well-Being—The Role of Task Interdependence

4.2.1 Introduction

Research beyond the domain of exercise psychology has already shown that the degree of interdependence as a central characteristic of tasks has an influence on social feelings and the quality of social interaction. While competition leads to an increase in aggression and hostility (even in winners; Anderson and Morrow 1995; Nelson et al. 1969), cooperation leads to interpersonal attraction (Garibaldi 1979) and facilitates the establishment of interpersonal relationships (Deutsch 1973).

Therefore, the aim of this study is to examine the influence of interdependence during an exercise task on social well-being as a consequence of need satisfaction. We hypothesize cooperative exercise tasks to lead to higher relatedness when compared to competitive tasks.

4.2.2 Method

Participants The sample consisted of 30 persons (11 females, 19 males) ranging from 19 to 33 years of age ($M = 22.27$, $SD = 2.97$). The participants were students of the local sport university, each of whom consented to participate in the study voluntary.

Treatment Participants were assembled into dyads, and each dyad was randomly assigned to one of three conditions. These conditions differed in the kind of interdependence the dyads were allocated to while completing a table tennis task. In the competitive condition, the two persons of each dyad were instructed to play table tennis against each other (negative interdependence). In more detail, the participants were encouraged to win as many rallies as possible in order to gain a maximum difference between won points and lost points compared to their opponent. Conversely, in the cooperative condition, the two persons of each dyad were instructed to play table tennis not against each other but supportively. Specifically, they were told to play each rally as long as possible (positive interdependence). Finally, in the individualized condition, the two persons of each dyad were instructed to play table tennis on their own (no interdependence) by hitting the ball against a glass panel which was positioned behind the table, in such a way that the ball bounces back on the table without hitting the floor. The aim of this condition was to make no error in each rally for as long as possible. Thus, although the conditions differed in terms of interdependence, each condition reflected a social situation (the presence of another person) in a similar sport setting (table tennis task).

Measures *Social well-being* Social well-being was measured using an adapted version of the German "Skala zum Betreuungsbefinden" (SBebe; Kleinert 2014). The SBebe was developed as a short form of a 12-item adjective list of social well-being (SOWEAL, Kleinert 2012), being based on self-determination theory (Deci and Ryan 2000). The SBebe consists of three bipolar items asking the participants to indicate their feelings of autonomy ("How autonomous did you feel?"), competence ("How competent did you feel?") and relatedness ("How related did you feel?") during the preceding table tennis task. For each of the three items, the response options range from 1 to 10, with both extremes being described by a triad of adjectives: autonomy (1 = heteronomous, dependent, forced; 10 = self-determined, voluntary, independent), competence (1 = overwhelmed, incapable, underestimated; 10 = capable, positively stressed, valuable) and relatedness (1 = excluded, disliked, misunderstood; 10 = recognized, considered, liked).

Interpersonal closeness Interpersonal closeness between the study participants of a dyad was measured using the "Inclusion of Other in the Self" scale (IOS; Aron et al. 1992). The scale asks the participants to indicate the closeness of their relationship to the other person by marking one of six Venn-like diagrams ("Please circle the picture below which best describes your relationship"). The diagrams consist of two circles which represent the asked person and the other person, with the variation in overlap representing different levels of closeness, ranging from no overlap (=1) to a nearly complete overlap (=6).

Procedure After gaining ethical consent for the study from the responsible university board, participants were recruited from the local campus. Potential participants were told that the aim of the study was to examine the influence of several sport activities on well-being. Persons willing to participate were assembled to dyads. After welcoming a dyad to the laboratory, the two participants were asked to fill in a questionnaire to collect demographic information. Subsequently, the dyad was randomly allocated to one of the three experimental conditions. Each dyad had to perform their assigned table tennis task for 10 min. Following the table tennis task, the participants received a paper-pencil questionnaire containing the measures of social well-being and interpersonal closeness. Upon completion of the data collection, participants were informed of the detailed study purpose and dismissed.

Data Analysis Data were analyzed using IBM SPSS Statistics 22.0. The analysis comprised descriptive and inferential examination of the study variables. In terms of inferential statistics, firstly, a one-way MANOVA was computed for social well-being. In order to investigate significant MANOVA effects, separate one-way ANOVAs and single comparisons using Bonferroni-adjusted post hoc tests were computed. Secondly, a one-way ANOVA was computed for interpersonal closeness, being followed by post hoc tests using Bonferroni adjustment. The overall significance level was set at $\alpha = 0.05$.

4.2.3 Results

In total, the sample had a high social well-being (autonomy $M = 7.23$, $SD = 1.96$; competence $M = 6.97$, $SD = 2.44$; relatedness $M = 8.17$, $SD = 1.21$). Particularly, the participants reported feeling highly related. However, social well-being varied strongly between participants, especially in terms of autonomy ($Min = 2$, $Max = 10$) and competence ($Min = 3$, $Max = 10$). Regarding interpersonal closeness, the sample reported neither feeling very close nor feeling very distant to the partner in general ($M = 3.43$, $SD = 1.72$). But, again, the individual level of felt interpersonal closeness varied strongly ($Min = 1$, $Max = 7$).

The multivariate examination of social well-being showed a significant, large effect of the treatment on social well-being, $F(6, 62) = 2.59$, $p = 0.028$, $\eta^2 = 0.23$. Three subsequent, separate one-way ANOVAs showed that the three experimental conditions differed only significantly with regard to relatedness, $F(2, 27) = 4.23$, $p = 0.025$, $\eta^2 = 0.24$. Bonferroni-adjusted post hoc tests revealed that the cooperative condition felt more related than the individualized condition (see Fig. 4.1 for descriptives), $p = 0.024$. The competitive condition neither differed from the cooperative ($p = 0.943$) nor from the individualized ($p = 0.228$) condition. In terms of autonomy and competence, the ANOVAs were not significant, $F(2, 27) = 2.97$, $p = 0.068$, $\eta^2 = 0.18$ and $F(2, 27) = 0.49$, $p = 0.618$, $\eta^2 = 0.04$. Post hoc tests

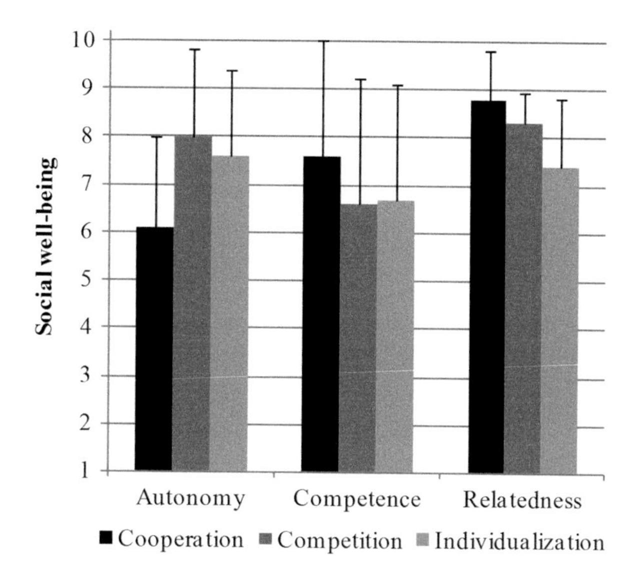

Fig. 4.1 Descriptive statistics of the experimental groups regarding social well-being

showed that the cooperative condition had at least the tendency of lower autonomy when compared to the competitive condition (see Fig. 4.1), $p = 0.086$. The univariate examination of interpersonal closeness did not find any group differences between the three experimental conditions, $F(2, 27) = 0.57$, $p = 0.571$, $\eta^2 = 0.04$.

4.2.4 Discussion

The aim of the study was to examine the influence of task interdependence during exercise activities on social well-being. Contrary to our assumption, the hypothesis that cooperative activities lead to higher feelings of relatedness than competitive activities could not be confirmed. However, despite this finding, the cooperative but not the competitive condition differed significantly from the individualized activity being conducted in a social setting in terms of relatedness. Thus, with regard to relatedness, the cooperative condition has a more beneficial effect than the competitive condition when compared to individualization. This is also reflected by the fact that, when regarded descriptively, relatedness was highest in the cooperative condition. This can be explained by the fact that the participants in the individualized condition were not immediately integrated into a task with another person. Furthermore, this can be related back to the fact that performing a task with a common goal (i.e. cooperating) leads to feelings of being considered and integrated.

As far as competition is concerned, there was no significant difference to the individualized condition or to the cooperation condition. Obviously, competition has an intermediate position when regarding the experimental effect. This may be due to the fact that competition situations contain both a potential for negative effect on social well-being ("the other is an opponent and I want to beat him/her") and a positive effect on social well-being ("competition is a game and we play this game together"). Moreover, there are two main reasons for the missing of an overall negative effect of the competitive condition when being compared to the cooperative condition. Firstly, the competition was not associated with any external consequences. In other words, winning or losing the competition did not lead to rewards or punishment, respectively. Secondly, it can be assumed that the participants liked the competition and appraised them as a joyful activity as they were all sport students being faced with smaller competitions every day. For both reasons, the typical feeling of having an opponent who thwarts one's goals and radiates some kind of hostile feelings might not have actually occurred.

Therefore, performing an exercise activity directly in interaction with another person (either with a cooperating person or with a competing person with the competition having no external consequences) is not sufficient per se to gain an optimal effect with regard to relatedness—it is, after all, still the cooperation leading to the highest relatedness.

Despite this positive side of interaction (particular cooperative), the tendency of lower levels of autonomous feelings in the cooperative activity when being compared to the individualized activity can be interpreted as a hint for possible negative effects of social interactions on well-being as well. The participants in the individualized condition had a free choice of when, where or how to hit the ball. Conversely, in the cooperative condition, the ball had to be played appropriately for the partner, and, in consequence, the participants had a lower degree of freedom. Hence, depending on the given conditions, interacting has a potential for reducing feelings of autonomy.

The aforementioned aspects of missing external consequences in the competitive condition and a restricted degree of freedom in the cooperative condition warrant further research. Future studies should involve negative (e.g. losing money) and/or positive (e.g. winning money) consequences in the competitive condition. Moreover, they should investigate cooperative, competitive and individualized tasks not only having comparable demands but also similar degrees of freedom. Nevertheless, our study results already indicate positive effects of cooperation on the relatedness aspect of social well-being.

4.3 Study 2: Reduction of Loneliness—The Role of Group Identification Within the Exercise Context

4.3.1 Introduction

Loneliness is "the unpleasant experience that occurs when a person's network of social relations is deficient in some important way" (Perlman and Peplau 1981, p. 31). At its core, it consists of feeling of emptiness and being left alone, resulting from a qualitative lack in the social relationships of an individual (De Jong Gierveld

1987). Hence, loneliness is a subjective experience and cannot be described objectively, in contrast to social isolation which is the objective state of having no one around (De Jong Gierveld et al. 2006).

Loneliness is associated with several negative health consequences. On the one hand, it leads to mental health problems as lonely people suffer from negative affect (e.g. depressive disorders Cacioppo et al. 2006). On the other hand, loneliness leads to physical health problems such as heart disease (see Lauder et al. 2006 for an overview). These concerns raise the question of how loneliness can be reduced. Therefore, the purpose of this study is to examine the conditions under which exercise can contribute to reduced feelings of loneliness.

Studies investigating the relationship between loneliness and sport or exercise can be categorized into studies examining the effect of controlled interventions, longitudinal studies without controlled interventions and cross-sectional studies. To date, there are four intervention studies (Hopman-Rock and Westhoff 2002; Kahlbaugh et al. 2011; McAuley et al. 2000; Savikko et al. 2010) examining and finding a loneliness-reducing effect of exercise programs on loneliness. This beneficial effect could be found irrespective of how the intervention programs were designed in detail. While each program consisted of exercising in a group, they differed with regard to their content and their duration. For example, they included specific, predefined exercise like aerobic exercise (McAuley et al. 2000) or more unspecific exercise with varying contents such as dancing, swimming or walking within one program (Savikko et al. 2010). The program periods were between 10 weeks (Kahlbaugh et al. 2011) and 8 months (Hopman-Rock and Westhoff 2002) with sessions held between one (Kahlbaugh et al. 2011; Savikko et al. 2010) and three times (McAuley et al. 2000) per week for a duration of 40 min (McAuley et al. 2000) up to 6 h (Savikko et al. 2010) each.

However, exercise is not associated with low feelings of loneliness per se. Only some (e.g. Page et al. 2003; Toepoel 2013), but not all (e.g. Cacioppo et al. 2002; Lauder et al. 2006), of the existing cross-sectional studies and none of the existing longitudinal (i.e. prospective) studies (Bohnert et al. 2013; Findlay and Coplan 2008) found a direct negative association between exercise-related measures and loneliness. Thus, potential mediators and moderators within this relationship have to be considered to clarify the conditions under which exercise actually leads to low feelings of loneliness.

In general, a potential beneficial effect of exercise on loneliness can be explained via the discrepancy model of loneliness. According to this model loneliness is resulting from a negative discrepancy between the degree of relatedness an individual needs in his or her life and the degree of relatedness he or she actually perceives. Thus, exercise activities enhancing relatedness by providing relationships with a positive quality should be favourable to reduce loneliness. In line with this, the intervention study from McAuley et al. (2000) found that the decrease in loneliness was related to an increase in perceived social support by the other course members. However, this is the only intervention study examining potential underlying mechanisms of the relationship between exercise and loneliness, although the role of social support was confirmed by a cross-sectional study from Taliaferro et al. (2010), reporting that perceived social support mediated the relationship between team sport participation and loneliness. Besides these consistent findings on social support, further mechanisms have to be identified.

Studies in sport spectator research found that spectators who identify with their team indicate less loneliness (Wann et al. 2003, 2011). Therefore, identification with an exercise group might be another explanation for the relationship between exercise and loneliness. Both social support and group identification are aspects of relationship quality by reflecting issues of group involvement, but they have distinct elements: Whereas social support by a group can be described as social provisions supplied by the group members to an individual (Wallston et al. 1983), group identification reflects the psychological relationship of an individual to his or her group (van Knippenberg et al. 2007).

Given these different orientations of social support and group identification in terms of relationship quality, it seems appropriate to examine the role of identification in the relationship between exercise and loneliness as well. Therefore, this study aims to test whether identification with an exercise group is related to loneliness. More specifically, it is hypothesized that individuals reporting higher levels of group identification with their exercise group feel less lonely than those reporting lower levels of identification.

4.3.2 Method

Participants The sample consisted of 192 persons (88 females, 104 males) ranging from 16 to 75 years of age ($M = 27.29$, $SD = 12.59$). In general, they took part in $M = 1.67$ ($SD = 0.78$) different exercise activities. All participants conducted physical activity in at least one exercise group (e.g. gymnastic group, health sport group). With regard to their main exercise group, participants were group members for an average of $M = 4.44$ ($SD = 5.45$) years. They practiced in these groups for an average of $M = 2.60$ ($SD = 1.41$) times per week and reported to meet at least some group members $M = 1.34$ ($SD = 1.59$) times per week additionally. Each of them indicated to participate in the study voluntary.

Measures *Loneliness* Loneliness was assessed using the German version of the revised UCLA Loneliness Scale (Döring and Bortz 1993). It allows to compute a global index of loneliness consisting of 20 items (e.g. "I feel rejected."). The response options range from 1 (=do not agree at all) to 5 (=fully agree). The global index had a good internal consistency (Cronbach's $\alpha = 0.87$).

Group identification Identification with the main exercise group was assessed by a German scale developed by Zepp and Kleinert (2015). Based on Tajfel's (1978) construction of identification, it measures a cognitive (e.g. "I see this group as part of who I am"; Cronbach's $\alpha = 0.68$), evaluative (e.g. "Personally, it is very important to me to be part of this group"; Cronbach's $\alpha = 0.69$) and affective factor (e.g. "I like being with my fellow team members"; Cronbach's $\alpha = 0.85$) of identification using two, respectively, three items per factor. The response options ranged from 1 (=do not agree at all) to 7 (=fully agree).

Procedure After gaining ethical consent for the study from the responsible university board, participants were recruited via several sport providers (e.g. sport clubs and gyms). Participants filled in a questionnaire relating to the aforementioned sociodemographic factors and exercise habits as well as to loneliness and group identification. Data analysis comprised descriptive statistics and regression analysis. Loneliness was regressed on identification with the exercise group (block 3) while controlling for age (block 1) and exercise habits (block 2) with forced entry.

4.3.3 Results

Study participants scored low on feelings of loneliness ($M = 1.69$, $SD = 0.36$). All participants reported values in the lower half of the 5-point Likert response scale ($Min = 1.20$, $Max = 2.90$). Identification with the exercise group tended to be slightly high. Particularly, the affective category ($M = 5.95$, $SD = 0.79$) and the evaluative ($M = 5.20$, $SD = 1.06$) category of identification were high, whereas cognitive identification ($M = 4.20$, $SD = 1.21$) scored moderately.

Loneliness and the categories of identification correlated negatively (between $r = -0.11$, $p > 0.05$ and $r = -0.25$, $p < 0.01$). In other words, individuals with higher feelings of identification reported to feel less lonely. However, the correlations had small effects. Moreover, the correlation between loneliness and evaluative identification was not significant. The correlations among the identification categories were positive with large effects (between $r = 0.46$, $p < 0.05$ and $r = 0.60$, $p < 0.05$).

The regression analysis demonstrated a negative association between the affective category of identification and loneliness ($\beta = -0.330$, $p = 0.002$). Contrary, neither the cognitive category ($\beta = -0.050$, $p = 0.617$) nor the evaluative category ($\beta = 0.113$, $p = 0.326$) was found to be a significant predictor. With regard to the control variables, age was positive ($\beta = 0.195$, $p = 0.045$), and the number of exercise activities was negatively associated ($\beta = -0.245$, $p = 0.027$) with loneliness.

In sum, the final model [$F(9, 141) = 2.78$, $p = 0.005$] explained $R^2 = 0.15$ of the variance in loneliness. This is equivalent to a medium effect. In contrast, the first model containing only age [$F(1, 149) = 3.55$, $p = 0.061$] and the second model containing only the exercise habits in addition to age [$F(6, 144) = 1.68$, $p = 0.130$] accounted for $R^2 = 0.02$ ($f^2 = 0.02$) and $R^2 = 0.07$ ($f^2 = 0.08$) of the variance in loneliness.

4.3.4 Discussion

In order to investigate the mechanisms being responsible for a loneliness-reducing effect of exercise, the present study analyzed the relationship between identification with an exercise group and loneliness. In concordance with our hypothesis, at least affective identification with an exercise group was a negative predictor of loneliness. However, cognitive and evaluative identification could not be identified as predictors of loneliness.

Two approaches, relating to an individual's innate, basic psychological need to belong, might explain that the finding that individuals with higher affective identification with their exercise groups feel less lonely. Firstly, affective identification reflects the feeling of attraction towards the group (Henry et al. 1999) and is, therefore, an expression for a strong affective bonding. Such a bonding satisfies the need to belong (see Baumeister and Leary 1995), whereas the affective state which is related to this bonding is the consequence of need satisfaction. Moreover, this process of need satisfaction can, in turn, minimize the discrepancy between interpersonal relationships as they subjectively are and interpersonal relationships as the individual desires them to be. Secondly, the satisfaction of the need to belong is associated with positive feelings (Deci and Ryan 2000), which might—in a more unspecific way—counteract or compensate for the negative feelings associated with loneliness.

The conceptualization of the cognitive and the evaluative category of identification as more rational and less emotional, in comparison to the affective category, might explain why both are less associated with loneliness. Thus, by reflecting the strength of categorizing oneself as a group member and by reflecting the importance of group membership, both do not refer to immediate feelings of social well-being within a group. In contrast, loneliness is a subjective experience, which, in terms of group memberships, apparently seems to depend more on irrational, immediate feelings of interpersonal attraction. Therefore, affective identification might be more associated with loneliness than both cognitive and evaluative identifications are.

However, this does not necessarily imply that the cognitive and the evaluative categories are entirely unrelated to loneliness. Both add to the social identity of a person and, therefore, can contribute to a broader sense of feeling societal integration and to a reduction in loneliness. An indicator of this assumption is the negative correlation between cognitive identification and loneliness. However, as both cognitive and evaluative identifications correlate with the affective category, this shared variance could be a methodological explanation as to why, finally, only the affective category emerged as a significant predictor.

According to our results, it is also important to consider the role of age and number of exercise activities when examining the physical activity and loneliness relationship. This study showed a positive correlation between age and loneliness, which can be explained by the fact that older adults have a higher prevalence of loneliness in general due to changing life circumstances (e.g. less qualitatively high contact to significant others, such as relatives and friends, due to cases of death or growing immobility; Dykstra 2009). In terms of number of exercise activities, it might be that a higher number is accompanied by memberships to multiple different sport groups. This, in turn, could contribute to less loneliness as a high diversity in relationships has a beneficial influence on loneliness (see De Jong Gierveld et al. 2006). In contrast, social monotony (i.e. low diversity in relationships) is associated with diminished social identities, perhaps contributing to higher feelings of loneliness. However, given the results of our regression model, the quality of relationships (i.e. affective identification) was stronger related to reduced loneliness than the quantity of relationships (i.e. frequency of exercise).

Taken together, identification and the control variables of age and exercise aspects accounted for 15 % of the variance in loneliness. Although this corresponds to a medium effect size (see Cohen 1992), this restricted explanation of variance can be explained by the fact that loneliness is a global construct in the sense of Vallerand (1997). As such, loneliness is influenced by several contexts, of which exercise (or physical activity in general) is one. Therefore, to improve predictions of loneliness (or even to compare the influence of different contexts), further contexts (e.g. family, work, etc.) should be taken into account.

The results are limited in terms of the study design. As the relationship between identification and loneliness was identified through a cross-sectional design, the causal relationship between these variables cannot be established. Therefore, future study designs should be longitudinal or even experimental.

In conclusion, the results of this study indicate that it is affective identification in particular that holds a significant role in an individual's level of loneliness. Therefore, exercise instructors should pay attention to the development of identification in a course group. For lonely individuals this provides a twofold benefit: Firstly, physical activity and exercising are domains of everyday life and, therefore, would be easily accessible in contrast to, for example, social cognition interventions (which are proposed by Masi et al. 2011). Secondly, physical activity holds an inherent additional value due to its potential for alleviating health problems associated with loneliness like depression (see Biddle and Mutrie 2008).

4.4 Study 3: Reduction of Aggressive Feelings—The Role of Movement Type and Task Interdependence

4.4.1 Introduction

Exercise activities are typically regarded as appropriate methods to reduce aggression and associated feelings (Jarvis 2006). However, the actual state of research leads to the conclusion that the relationship between exercise and aggressive feelings such as testiness or anger is unclear (Williams and Gill 2000). Whereas some researchers have found exercise related with low aggression (Kreager 2007), others have found a non-significant (Begg et al. 1996) or even an inverse relationship (Mason and Wilson 1988).

As already stated by Kleinert and Kleinknecht (2012), such inconsistent findings necessitate the consideration of potential moderators of the exercise-aggression link. For example, these moderators could be found in terms of the target group an activity is designed for, social aspects of the activity and the movement type of the activity. Thus, the present study aims to investigate selected moderators that assumably explain when and why aggressive feelings are enhanced or reduced through exercise.

With regard to the target group, studies indicate that sex acts as a moderator for the exercise-aggression link at least for adolescents. Several studies showed that positive relationships between exercise and aggression are more likely for females. For example, Mistretta (2006) found that the relationship between aggression-related measures and sport was positive for girls but not for boys. Similarly, Begg et al. (1996) found that the probability of delinquency was heightened twofold for both boys doing sports and 3.2-fold for girls doing sports. To sum up, it can be assumed that sport or exercise activities are more disinhibiting with regard to aggression for girls than for boys, for example, for biological (e.g. hormones) or for psychosocial (e.g. self-concept, sex role) reasons.

Furthermore, studies indicate that movement type might act as a moderator. Some sport activities are typically assumed to increase aggressiveness, in particular combat sports such as boxing, kickboxing and wrestling (Jarvis 2006). Beginning such combat sports is associated with an increase in antisocial behaviour, whereas quitting these sports leads to a decrease in antisocial behaviour (Endresen and Olweus 2005). One argument could be that aggressive movements themselves (e.g. kicking or hitting the opponent) involved in these sports have an influence on the amount of aggressiveness. Other authors argue that social norms lead to higher aggression in some activities. As Maxwell et al. (2009) summarize, it can be assumed that aggression as a legitimate behaviour is more likely in contact and collision sports due to inherent norms accepting this behaviour. In addition to the type of activity undertaken, levels of aggressiveness may also vary depending on the style in which a specific activity is performed. For example, modern martial arts training leads to higher scores in aggressiveness than more traditional styles (Nosanchuk and MacNeil 1989; Trulson 1986).

As far as social conditions are concerned, aggression research in general focuses specifically on the effects of competition on aggressiveness. According to Berkowitz (1993), it can be assumed that competition may lead to an increase in aggressiveness in general. As opponents interfere conversely with the achievement of individual goals, this could cause frustration and, consequently, lead to higher aggression. This assumption is supported by results of classic experiments (Sherif et al. 1961) but also by more recent research (Adachi and Willoughby 2013; Breuer et al. 2014).

To sum up, recent research on aggression reduction through exercise suggests that a reduction in aggression seems to depend on sex and movement type. However, as Kleinert and Kleinknecht (2012) remarked before, recent research lacks experimental designs that allow for unambiguous causal statements. Additionally, to our knowledge, there is no study explicitly focusing on the exercise-aggression link with regard to the moderating role of competition. For these reasons, we aimed to test two hypotheses experimentally. Firstly, we assume that the decrease in aggressive feelings (e.g. anger, testiness) is stronger with movements that are not similar to aggressive actions (e.g. neutral movements such as rowing) compared to movements with an aggressive connotation (e.g. boxing, hitting). Based on Berkowitz (1993), our second assumption is that the presence of an opponent restricts the reducing effect of exercise on aggressive feelings; thus, we hypothesize that aggressiveness scores will be less reduced (i.e. higher) when completing a task against an opponent than when completing an individual task.

4.4.2 Method

Participants The sample consisted of 39 persons (17 females and 22 males) ranging from 20 to 26 years of age ($M = 22.69$, $SD = 1.67$). The participants were students of the local sport university. Each of them indicated to participate in the study voluntary.

Treatment Participants were allocated to one of four different treatment groups. The treatments differed in terms of the factor "movement type" and the factor "opponent". The factor "movement type" implied either rowing or combat sport. Participants in the rowing condition rowed on an ergometer. Participants in the combat sport condition had to perform combat sport in the two treatment phases. Participants doing combat sport had to punch a boxing bag or to punch against an opponent using a padded foam bat (bataka bat), depending on whether they had to accomplish a single or a competitive task on the factor "competition". The factor "competition" implied either a single or a competitive task. In the single task condition, participants had to perform a solo movement. In the competitive rowing condition, participants competed against an opponent. Participants doing competitive combat sports were instructed to fight against an opponent. The opponent in each condition was an assistant of the examiner.

Measures Aggressive feelings (e.g. anger, testiness) were measured using an item of the German eight-item adjective list (SBS; Hackfort and Schlattmann 1995). The item consists of three similar adjectives ("angry/testy/displeased") asking the participants to indicate their momentary aggressive feelings. The response options range from 0 (is not true) through 5 (moderately true) to 10 (is very true).

Procedure After gaining ethical consent for the study from the responsible university board, participants were recruited from the local campus. Potential participants were told that the aim of the study was to examine whether self-perception differs between two activities. Actually, the first of the two activities served to induce aggressive feelings, whereas the second consisted either of the rowing or the combat sport exercise. In more detail, after welcoming the participants to the laboratory, the participants obtained an aggression induction via unfair and harsh feedback from the examiner's assistant in the course of the first activity, namely, several specific table tennis tasks (e.g. keeping the ball up 20 times in a row without dropping it on the floor). After completing these tasks successfully, participants received a paper-pencil questionnaire containing the aforementioned measures ($t1$). Subsequently, the participants were randomly allocated to one of the four treatment conditions. Each condition was separated into two phases containing the same treatments (first and second treatment phase), representing the second activity. After the first and after the second treatment phase, a paper-pencil questionnaire was delivered once again ($t2$, $t3$). Following the final data collection, participants were informed of the actual study purpose and dismissed.

Data analysis Data were analyzed using IBM SPSS Statistics 22.0. The analysis comprised descriptive and inferential examination of the study variables. In terms of inferential statistics, a separate 2 by 2 by 3 ANOVA with two twofold group factors (movement type and competition) and a threefold factor for repeated measures were computed. In order to investigate significant ANOVA effects, single comparisons using *t*-tests were computed. The single comparisons were manually Bonferroni adjusted. The overall significance level was set at $\alpha = 0.05$.

4.4.3 Results

Aggressive feelings changed significantly over time, $F(2, 70) = 6.99$, $p = 0.002$, $\eta^2 = 0.17$. As indicated by Bonferroni-adjusted single comparisons (corrected significance level, $p = 0.017$), aggressive feelings decreased in the course of the first treatment phase ($M_{t1} = 2.41$, $SD_{t1} = 2.61$; $M_{t2} = 1.41$, $SD_{t2} = 1.68$; $t(38) = 2.68$, $p = 0.011$, $d = 0.09$). There was no further significant change of aggressive feelings in the course of the second treatment phase ($M_{t3} = 1.21$, $SD_{t3} = 2.00$).

There was no significant main effect of movement type on aggressive feelings. However, time and movement type interacted significantly with regard to aggressive feelings, $F(2, 70) = 5.82$, $p = 0.005$, $\eta^2 = 0.14$. Specifically, only those participants in the rowing condition showed a decrease in aggressive feelings, whereas those doing combat sports stayed at the same level (ordinal interaction; see Fig. 4.2). This was indicated by six separate Bonferroni-adjusted single comparisons (three comparisons for each movement type; corrected significance level, $\alpha = 0.008$) that demonstrated a significant decrease from the aggression induction to the end of the first treatment phase [$t(19) = 3.41$, $p = 0.003$] and from the aggression induction to the end of the second treatment phase [$t(19) = 3.70$, $p = 0.002$] in the rowing condition only.

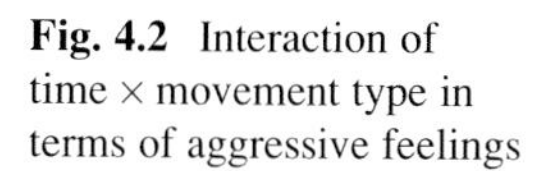

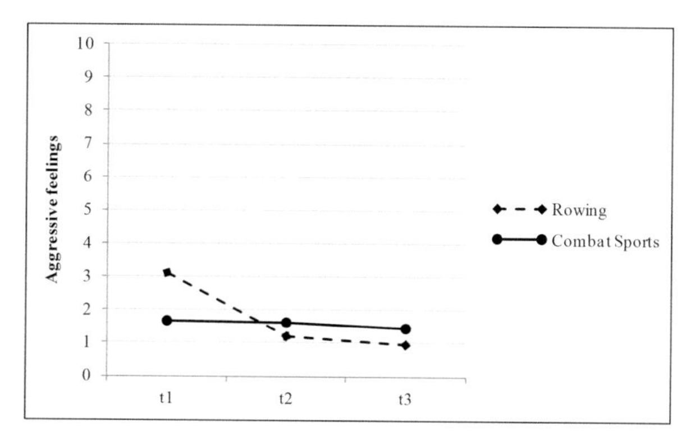

Fig. 4.2 Interaction of time × movement type in terms of aggressive feelings

4.4.4 Discussion

The study aimed to examine the conditions under which sport and exercise contribute to a reduction of aggressive feelings. It was found that not the social conditions (i.e. the absence or presence of an opponent) but the type of exercise activity (i.e. rowing or combat sport) influences the change in aggressive feelings.

Rowing led to a reduction in aggressive feelings, whereas combat sport did not. This reduction could be observed already within the first treatment phase; there was no additional reduction in aggressive feelings during the second exercise phase. As the aggressive feelings in the rowing exercise decreased from a low level (3.10 on an 11-point Likert scale) to a very low level (1.20), this change demonstrates that a reduction in aggressive feelings during the course of exercise is possible even at relatively low levels.

The fact that rowing may lead to a greater decrease in aggressive feelings compared to combat sport seems to support the popular view that martial arts hold an aggressive potential (Jarvis 2006). However, during the combat sport of this study, there was no increase in aggressive feelings. This is an ambiguity, reflecting the conflicting findings of the recent research in this area (see Williams and Gill 2000).

There was no difference in aggressive feelings between the two social conditions (i.e. the absence vs. presence of an opponent). It was assumed that the presence of a competitor would lead to an increase in aggressive feelings as the opponent would permanently interfere with the opportunity of an individual to achieve a goal, leading to frustration and, in consequence, to more aggressive feelings (Berkowitz 1993; Tjosvold 1998). With regard to the present experiment, it is possible that the goal importance was not high enough for the participants to provoke a suchlike effect. Moreover, it can be assumed that positive effects of an opponent on aggressive feelings need a minimum level of negative relationship; in other words, if an individual likes an opponent even if they are directly competing, this competition condition would not lead to aggressive feelings against this person.

There are two limitations of the present study. Firstly, the initial levels of aggressive feelings are rather low, suggesting a floor effect and possibly indicating that the induction was not strong enough to evoke intense aggressive feelings (this is particularly true for the combat sport group in our experiment). Secondly, it is debatable whether the movement types within the lab setting of the experiment are reflective of the actual exercise conditions under ecological conditions. These points limit the practical application of our results.

Given these limitations, future research should develop stronger aggression induction procedures to exaggerate the effects of this experiment (i.e. beneficial effect of rowing movements compared to combat sport movement). Moreover, as far as competition is concerned, the induction of negative relationships between competitors seems to be relevant to examine competition effects. Different forms of sport or exercise have to be investigated regarding their effect on each of these two components and their interaction in regard to the regulation of aggressive feelings.

4.5 General Discussion

Even if the three aforementioned studies had different aims, different designs and consider different psychological constructs, they have some aspects in common which should be discussed in this final section. First of all, all studies look on exercise settings in regard to social well-being. However, both exercise and social well-being had been operationalized in different ways. In terms of exercise, study 1 looked at table tennis in a situational, experimental paradigm, participants in study 2 conducted different forms of exercise in groups or sport clubs and study 3 took combat sport or rowing—again in an experimental setting—into account. Obviously, these different forms of exercise are hard to compare, since intensity, type of exercise behaviour or duration of exercise was different. Thus, a result stemming from one exercise setting is not easy to transfer into another exercise setting. However, to enhance the opportunity of transfer and generalization, this general discussion tries to identify principal mechanisms. That is, instead of discussing specific mechanisms taking into account specific forms of exercises, an abstract or conceptional level should be reached to enable the transfer of results. The same approach applies to the different forms of social well-being considered in the three studies. Specifically, actual feelings of relatedness or autonomy (study 1), rather stable experiences of loneliness (study 2), or state feelings of aggression (study 3) are different aspects of social well-being. Again, this discussion tries to find a global, comparable level based on these specific and different forms of social well-being.

Besides this basic problem of generalization and transfer of treatments (exercise) and outputs (social well-being), the three studies dealt with two forms of moderators: Two studies (study 1, study 3) considered effects of *competitive settings*, and two studies (study 1, study 2) examined *social relationship* as a moderator of the link between exercise and social well-being. Therefore, this general discussion is structured in basically three parts, one on social relationship as a moderator of the exercise-well-being link and one on competition settings as a moderator in this concern. A third and last section deals with the implications for Space science, that is, the transfer of the discussed moderating processes in regard to social well-being of members in work group in Space labs.

4.5.1 Social Relationship as a Moderator

The definition and description of social well-being, which was developed in the introductory part of the chapter, pointed already to social relationship as a dominant characteristic of social well-being. Specifically, social interaction was defined as the basic source of social well-being, whereas appraisal of social interactions was assumed to moderate this relationship. Moreover, according to Larson (1993) and Keyes (1998), different aspects of appraisal were described: An individual evaluates the way of social adjustment, social support and social integration in the course or after social interaction. In these aspects of appraisal and evaluation, the social relationship is both a predictor and a consequence of appraisals. For instance, if a person likes the other (interacting) person, it is more likely that social interaction is evaluated as supportive or that one feels socially integrated, which leads to (social) well-being. Conversely, the perception of support or the feeling of being integrated does not only lead to social well-being but also to a positive attitude or feeling towards the other person (who is supporting or integrating), which is a positive development of a social relationship.

In study 1, opportunities for social integration and social support were experimentally induced by defining a cooperative task. Such tasks do not necessarily lead to a better social relationship; however, if the partners or group members recognize their interdependence in reaching a common goal, it may enhance the quality of relationship. Such consequences of common goals are established through social identity theory (Tajfel 1978). However, if the common goal or interdependent task is not enough to build a positive social relationship (for instance, due to negative pre-experiences between the partners), this would not only affect the quality of the common performance but also the social well-being of the partners. The result of study 1 gives evidence for the potential of cooperative exercise tasks for the positive development of social relationships (and positive social well-being) and confirms former research and theories (Deutsch 1973; Garibaldi 1979).

Social identity was already mentioned as a correlate of social relationship. Thus, mechanisms of social identity theory may also be discussed in relation to the development of positive social relationships in partner or group exercises. The cooperative task or the common goal was already stated as one of these identity mechanisms that lead to a stronger social identity, a positive relationship and higher extents of social well-being. However, exercising together with a partner or in a group could be a necessary, but not in itself sufficient, condition to enhance social identity and social well-being. Instead, it is important that both partners strongly identify with such shared activities and common goals. In principle, group members have to identify with the group as a whole, as activities, attitudes or goals are part of group characteristics. Thus, study 2 was designed to show whether group identification (as a moderator of the exercise-well-being link) enhances social well-being. In the regression model, group identification was the strongest predictor of social well-being (considered in this study to be the absence of loneliness) which confirms the important position of identification as a process within the development of social relationships.

Moreover, study 2 may suggest that identification and social identity are linked to the satisfaction of the need for belonging (Baumeister and Leary 1995) and relatedness (Deci and Ryan 2008). In our view, social well-being is a function of the degree to which social interactions in a certain social context satisfy basic psychological needs. It can be assumed that the extent of identification with the partner or the group (i.e. the extent of social identity) is positively correlated with the satisfaction of basic social needs. Thus, social identity is not only a correlate for a positive social relationship but also a condition for the satisfaction of social needs (i.e. relatedness, belonging). In this view, social well-being is simply the subjective experience of being socially satisfied.

Finally, our studies did not address the specific function or social role that an individual fills during social interactions in partner or group exercises. Such consideration would enable a closer insight into mechanisms by which relationships enhance or reduce social well-being. For instance, whether a group member feels a sort of social contribution (Keyes 1998) depends on his or her role (e.g. as a group leader). In this regard, the question of group similarity or group complementarity as conditions of identification and social well-being is also necessary (Zepp and Kleinert 2015). Such processes have to be addressed in future studies.

4.5.2 Competition as a Moderator

Considering the definition and description of social well-being, from a theoretical point of view, competition in immediate social interaction as a situational characteristic of an exercise task can be assumed to influence the appraisal of social interactions and, consequently, social well-being. In other words, as long as competition takes place in immediate social interaction, the course of a competitive exercise can be assumed to affect social well-being by the evaluation of one's social adjustment, social support and social integration, which are aspects of appraisal of social interaction (Keyes 1998; Larson 1993). As there are a limited number of winners in a competition, this circumstance is likely to influence how people feel adjusted, supported and integrated. In study 1, competition was experimentally created by a competitive table tennis task. In study 3, competition was experimentally induced through a competitive rowing task and a competitive combat sport task.

In both studies, competition had no specific effects on momentary social well-being. Irrespective of whether social well-being was measured in general (study 1) or in the form of specific social feelings (i.e. aggressive feelings; study 3), competition led neither to high nor to low social well-being. This might be explained by the fact that each aspect of appraisal of social interaction can have a positive extent or a negative extent during a competitive exercise despite the individuals involved being opponents. In terms of social adjustment, some individuals might, for example, show the same behaviours in the competitive situation (e.g. no cheering after successful actions) as their opponent and feel some kind of adjustment (positive extent), whereas others might show a contrary behaviour (e.g. being more offensive in gestures; negative extent) compared to their opponent. Regarding social support, it is usual that individuals receive little or no assistance from their opponents. However, in some cases, the opponent might help the individual to maintain the activity when it is joyful. In a similar way, social integration is usually low as the competitive task does not have to be completed together. Nevertheless, grasping competition as a game and playing this game together can potentially lead to feeling socially integrated. In consequence, competition does not seem to be necessarily unfavourable for social well-being.

Whether the appraisal of the social interaction during competition is positive or negative depends on both the preceding relationship of the involved persons and the actual consequences of the competition. In terms of the preceding relationship, the positive appraisal might only occur if the relationship is already positive before the competition takes place. If the relationship is negative and associated with feelings of anger towards the opponent prior to the competition, it might lead to negative appraisal. In terms of the consequences of the competition, rewards for winning or punishment for losing the competition might lead to negative appraisal of, in particular, social support and social integration as it leads to a fight for resources, thus increasing the valence of the competition.

It can be assumed that a momentary, positive appraisal of social interaction contributes to the development of a social identity if the activity is conducted regularly, despite the underlying task being competitive in nature. By sharing a common task with another person and by enjoying the task, people strongly identify with such shared activities and develop a social identity. This development of a social identity might, in turn, lead to a more stable relationship and more stable social well-being in the long term.

4.5.3 Implications for Manned Spaceflight and Space Science

The moderating role of social relationships and competition on the influence of exercise activities on social well-being and ill-being, respectively, is not only relevant for life on earth but also for life in Space labs. As it is well established that manned spaceflight needs countermeasures for social ill-being (e.g. Kanas and Manzey 2008; Manzey 2004; Manzey et al. 1995), the questions arise on how exercise countermeasures to social ill-being could be designed for Space labs. In this regard, it has to be considered whether already existing countermeasures to problems of physical health (e.g. steppers; Streeper et al. 2011) can be designed as countermeasures to social ill-being as well.

From the current state of research, in general, activities aiming at increasing social well-being and decreasing social ill-being could be cooperative or competitive in nature. However, with regard to competitive activities, these should only be conducted if the relationship is already positive before the competition takes place. Moreover, winning or losing the competition should be free of serious consequences which could be particularly a problem in Space as the resources in Space are limited and could lead to enviousness.

The exercise activities used in our experiments seem to be principally transferable to Space labs with some adaptations. Certainly, it is not possible to play table tennis in a Space lab, but already existing facilities like ergometers could be used for synchronous tasks as one example of cooperation. In contrast, activities which are perceived to be inherently unsocial (e.g. jogging alone; Vitulli and DePace 1992) should be avoided when aiming at an increase of social well-being. However, future studies still need to test which of the existing facilities could actually be used for an increase in social well-being. In particular, the external validity of these facilities should be verified in confinement studies.

References

Adachi PJC, Willoughby T (2013) Demolishing the competition: the longitudinal link between competitive video games, competitive gambling, and aggression. J Youth Adolesc 42:1090–1104

Anderson CA, Morrow M (1995) Competitive aggression without interaction: effects of competitive versus cooperative instructions on aggressive behavior in video games. Pers Soc Psychol Bull 21(10):1020–1030

Aron A, Aron EN, Smollan D (1992) Inclusion of other in the self scale and the structure of interpersonal closeness. J Pers Soc Psychol 63:596–612

Barrick MR, Stewart GL, Neubert MJ, Mount MK (1998) Relating member ability and personality to work-team processes and team effectiveness. J Appl Psychol 83:377–391

Baumeister RF, Leary MR (1995) The need to belong. Desire for interpersonal attachments as a fundamental human motivation. Psychol Bull 117(3):497–529

Begg DJ, Langley JD, Moffitt T, Marshall SW (1996) Sport and delinquency: an examination of the deterrence hypothesis in a longitudinal study. Br J Sports Med 30(4):335–341

Berkowitz L (1993) Aggression. Its causes, consequences, and control. McGraw-Hill, New York

Biddle S (1995) Exercise and psychosocial health. Res Q Exerc Sport 66(4):292–297

Biddle SJH, Mutrie N (2008) Psychology of physical activity: determinants, well-being, and interventions. Routledge, London

Bohnert AM, Aikins JW, Arola NT (2013) Regrouping: organized activity involvement and social adjustment across the transition to High School. New Dir Child Adolesc Dev 2013(140):57–75

Breuer J, Scharkow M, Quandt T (2014) Sore losers? A reexamination of the frustration–aggression hypothesis for colocated video game play. Psychol Pop Media Cult 42:126–137
Cacioppo JT, Hawkley LC, Crawford E, Ernst JM, Burleson MH, Kowalewski RB, Malarkey WB, van Cauter E, Berntson GG (2002) Loneliness and health: potential mechanisms. Psychosom Med 64:407–417
Cacioppo JT, Hughes ME, Waite LJ, Hawkley LC, Thisted RA (2006) Loneliness as a specific risk factor for depressive symptoms: cross-sectional and longitudinal analyses. Psychol Aging 21 (1):140–151
Cohen J (1992) A power primer. Psychol Bull 112:155–159
De Jong Gierveld J (1987) Developing and testing a model of loneliness. J Pers Soc Psychol 53 (1):119–128
De Jong Gierveld J, van Tilburg T, Dykstra PA (2006) Loneliness and social isolation. In: Vangelisti AL, Perlman D (eds) The Cambridge handbook of personal relationships. Cambridge University Press, Cambridge, pp 485–500
Deci EL, Ryan RM (2000) The "what" and "why" of goal pursuits. Human needs and the self-determination of behavior. Psychol Inq 11(4):227–268
Deci EL, Ryan RM (2008) Hedonia, eudaimonia, and well-being. An introduction. J Happiness Stud 9(1):1–11
Deci EL, Eghrari H, Patrick B, Leone DR (1994) Facilitating internalization: the self-determination theory perspective. J Pers 62(1):119–142
Deutsch M (1973) Conflicts: productive and destructive. In: Jandt FE (ed) Conflict resolution through communication. Harper & Row, New York
Döring N, Bortz J (1993) Psychometrische Einsamkeitsforschung. Deutsche Neukonstruktion der UCLA Loneliness Scale. Diagnostica 39(3):224–239
Duda JL, Balaguer I (2007) Coach-created motivational climate. In: Jowett S, Lavallee D (eds) Social psychology in sport. Human Kinetics, Champaign, IL, pp 117–130
Dykstra PA (2009) Older adult loneliness: myths and realities. Eur J Ageing 6(2):91–100
Endresen IM, Olweus D (2005) Participation in power sports and antisocial involvement in preadolescent and adolescent boys. J Child Psychol Psychiatry 46(5):468–478
Findlay LC, Coplan RJ (2008) Come out and play: Shyness in childhood and the benefits of organized sports participation. Can J Behav Sci 40(3):153–161
Fransen K, Coffee P, Vanbeselaere N, Slater M, de Cuyper B, Boen F (2014) The impact of athlete leaders on team members' team outcome confidence: a test of mediation by team identification and collective efficacy. Sport Psychol 28(4):347–360
Garibaldi AM (1979) Affective contributions of cooperative and group goal structures. J Educ Psychol 71(6):788–794
Glenister D (1996) Exercise and mental health: a review. J R Soc Promot Health 116(1):7–13
Hackfort D, Schlattmann A (1995) Die Stimmungs- und Befindensskalen (SBS). Institut für Sportwissenschaft und Sport, Neubiberg
Henry KB, Arrow H, Carini B (1999) A tripartite model of group identification: theory and measurement. Small Group Res 30(5):558–581
Hopman-Rock M, Westhoff MH (2002) Development and evaluation of "Aging Well and Healthily": a health-education and exercise program for community-living older adults. J Aging Phys Act 10:364–381
Jarvis M (2006) Sport psychology. A student's handbook. Routledge, London
Kahlbaugh PE, Sperandio AJ, Carlson AL, Hauselt J (2011) Effects of playing Wii on well-being in the elderly: physical activity, loneliness, and mood. Act Adapt Aging 35(4):331–344
Kanas N (2005) Interpersonal issues in space: Shuttle/mir and beyond. Aviat Space Environ Med 76(6):126–134
Kanas N, Manzey D (2008) Space psychology and psychiatry. Springer, Dordrecht
Kanas N, Sandal G, Boyd J, Gushin VI, Manzey D, North R, Leon G, Suedfeld P, Bishop S, Fiedler E, Inoue N, Johannes B, Kealey D, Kraft N, Matsuzaki I, Musson D, Palinkas L, Salnitskiy V, Sipes W, Stuster J, Wang J (2009) Psychology and culture during long-duration space missions. Acta Astronaut 64(7–8):659–677
Keyes CLM (1998) Social well-being. Soc Psychol Q 61:121–140

Keyes CLM, Shmotkin D, Ryff CD (2002) Optimizing well-being: the empirical encounter of two traditions. J Pers Soc Psychol 82(6):1007–1022
Kleinert J (2012) Social well-being as need satisfaction in social interaction. A social well-being adjective list. In: Martens T, Vollmeyer R, Rakoczy K (eds) Motivation in all spheres of life. International conference on motivation, Program & Abstracts, 28–30 Aug 2012, Frankfurt am Main. Pabst Science, Lengerich, p 123
Kleinert J (2014) Toolbox Beziehungsarbeit: Zur Beziehungsqualität in der sportpsychologischen Betreuung. Beitrag Qualitätssicherung in der Sportpsychologie. Sportverlag Strauß, Köln
Kleinert J, Kleinknecht C (2012) Sportliche Aktivität, Aggression und Gewalt. In: Fuchs R, Schlicht W (eds) Seelische Gesundheit und sportliche Aktivität. Hogrefe, Göttingen, pp 272–293
Kreager DA (2007) Unnecessary roughness? School sports, peer networks, and male adolescent violence. Am Sociol Rev 72(5):705–724
Larson JS (1993) The measurement of social well-being. Soc Indic Res 28(3):285–296
Lauder W, Mummery K, Jones M, Caperchione C (2006) A comparison of health behaviours in lonely and non-lonely populations. Psychol Health Med 11(2):233–245
Lewin K (1943) Defining the field at a given time. Psychol Rev 50(3):292–310
Manzey D (2004) Human missions to Mars: new psychological challenges and research issues. Acta Astronaut 55(3–9):781–790
Manzey D, Schiewe A, Fassbender C (1995) Psychological countermeasures for extended manned spaceflights. Acta Astronaut 35(4/5):339–361
Masi CM, Chen H, Hawkley LC, Cacioppo JT (2011) A meta-analysis of interventions to reduce loneliness. Pers Soc Psychol Rev 15(3):219–266
Mason G, Wilson P (1988) Sport, recreation and juvenile crime: an assessment of the impact of sport and recreation upon aboriginal and non-aboriginal youth offenders. Australian Institute of Criminology, Canberra
Maxwell J, Visek A, Moores E (2009) Anger and perceived legitimacy of aggression in male Hong Kong Chinese athletes: effects of type of sport and level of competition. Psychol Sports Exerc 10(2):289–296
McAuley E, Blissmer B, Marquez DX, Jerome GJ, Kramer AF, Katula J (2000) Social relations, physical activity and well-being in older adults. Prev Med 31(5):608–617
Mistretta LH (2006) Interscholastic athletic programs and juvenile delinquency in America's schools. Master Thesis, Washington
Nelson JD, Gelfand DM, Hartmann DP (1969) Children's aggression following competition and exposure to an aggressive model. Child Dev 40(4):1085–1097
Nosanchuk TA, MacNeil ML (1989) Examination of the effects of traditional modern martial arts training on aggressiveness. Aggr Behav 15:153–159
Ntoumanis N, Vazou S, Duda JL (2007) Peer-created motivational climate. In: Jowett S, Lavallee D (eds) Social psychology in sport. Human Kinetics, Champaign, IL, pp 145–156
Page RM, Lee C, Miao N, Dearden K, Carolan A (2003) Physical activity and psychosocial discomfort among high school students in Taipei, Taiwan. Int Q Community Health Educ 22 (3):215–228
Paluska SA, Schwenk T (2000) Physical activity and mental health. Current concepts. Sports Med 29(3):167–180
Pennebaker JW (1982) The psychology of physical symptoms. Springer, New York, NY
Perlman D, Peplau LA (1981) Toward a social psychology of loneliness. In: Duck SW, Gilmour R (eds) Personal relationships. Personal relationships in disorder. Academic, London, pp 31–56
Reinboth M, Duda JL (2006) Perceived motivational climate, need satisfaction and indices of well-being in team sports. A longitudinal perspective. Psychol Sports Exerc 7:269–286
Ryan RM, Deci EL (2001) On happiness and human potentials. A review of research on hedonic and eudaimonic well-being. Annu Rev Psychol 52:141–166

Ryff CD, Dienberg Love G, Urry HL, Muller D, Rosenkranz MA, Friedman EM, Davidson RJ, Singer B (2006) Psychological well-being and ill-being: do they have distinct or mirrored biological correlates? Psychother Psychosom 75(2):85–95

Savikko N, Routasalo P, Tilvis R, Pitkälä K (2010) Psychosocial group rehabilitation for lonely older people. Favourable processes and mediating factors of the intervention leading to alleviated loneliness. Int J Older People Nurs 5(1):16–24

Schlicht W (1995) Wohlbefinden und Gesundheit durch Sport. Hofmann, Schorndorf

Sherif M, Harvej OJ, White BJ, Hood WR, Sherif CW (1961) Intergroup conflict and cooperation. The Robbers Cave experiment. University of Oklahoma Book Exchange, Norman

Simon B (2004) Identity in modern society. A social psychological perspective. Blackwell, Malden, MA

Streeper T, Cavanagh PR, Hanson AM, Carpenter RD, Saeed I, Kornak J, Frassetto L, Grodsinsky C, Funk J, Lee S, Spiering BA, Bloomberg J, Mulavara A, Sibonga J, Lang T (2011) Development of an integrated countermeasure device for use in long-duration spaceflight. Acta Astronaut 68(11–12):2029–2037

Tajfel H (1978) Social categorization, social identity and social comparison. In: Tajfel H (ed) Differentiation between social groups. Academic, London, pp 61–76

Taliaferro LA, Rienzo BA, Miller MD, Pigg RM, Dodd VJ (2010) Potential mediating pathways through which sports participation relates to reduced risk of suicidal ideation. Res Q Exerc Sport 81(3):328–339

Tjosvold D (1998) Cooperative and competitive goal approach to conflict: accomplishments and challenges. Appl Psychol 47(3):285–313

Toepoel V (2013) Ageing, leisure, and social connectedness: how could leisure help reduce social isolation of older people? Soc Indic Res 113(1):355–372

Trulson ME (1986) Martial arts training: a novel 'cure' for juvenile delinquency. Hum Relat 39 (12):1131–1140

Turner JC, Hogg MA, Oakes PJ, Reicher SD, Wetherell MS (eds) (1987) Rediscovering the social group. A self-categorization theory. Blackwell, Oxford

Vallerand RJ (1997) Toward a hierarchical model of intrinsic and extrinsic motivation. In: Zanna (ed) Advances in experimental social psychology. Acadamic, San Diego, pp 271–360

van Knippenberg D, Haslam SA, Platow MJ (2007) Unity through diversity: value-in-diversity beliefs, work group diversity, and group identification. Group Dyn Theory Res Pract 11 (3):207–222

Vansteenkiste M, Ryan RM (2013) On psychological growth and vulnerability: basic psychological need satisfaction and need frustration as a unifying principle. J Psychother Integr 23 (3):263–280

Vitulli WF, DePace AN (1992) Manifest reasons for jogging and for not jogging. Percept Mot Skills 75:111–114

Wallston BS, Alagna SW, DeVellis BM, DeVellis RF (1983) Social support and physical health. Health Psychol 2(4):367–391

Wann DL, Dimmock JA, Grove JR (2003) Generalizing the team identification-psychological health model to a different sport and culture: the case of Australian rules football. Group Dyn Theory Res Pract 7(4):289–296

Wann DL, Rogers K, Dooley K, Foley M (2011) Applying the team identification–social psychological health model to older sport fans. Int J Aging Hum Dev 72(4):303–315

Williams L, Gill DL (2000) Aggression and prosocial behavior. In: Gill DL (ed) Psychological dynamics of sport and exercise. Human Kinetics, Champaign, IL, pp 239–254

Zepp C, Kleinert J (2015) Symmetric and complementary fit based on prototypical attributes of soccer teams. Group Processes Intergroup Relat 18(4):557–572

Chapter 5
Neurocognitive and Neuro-affective Effects of Exercise

V. Abeln, T. Vogt, and S. Schneider

Abstract Spaceflight is known to include the risk for physiological deconditioning as well as mental impairments, endangering mission safety and success. While Space science primarily focussed on physical health in the past, mental health (i.e. cognitive performance and affective state) was widely neglected. Today, not only in Space but also exercise science, the promotion of exercise for mental health is on the rise. This chapter is dedicated to review the known cognitive and affective deficits in relation to spaceflight. It aims to provide information about the underlying neurophysiological mechanisms, thus discussing the effect and application of exercise as a countermeasure, and to illustrate the contribution of Space science to mental health for our society.

Keywords Human Space science • Exercise science • Psychophysiological state • Mental health • Brain activity • Exercise recommendation • Cognitive performance

---

V. Abeln (✉) • T. Vogt
Institute for Movement and Neurosciences, German Sport University Cologne, Am Sportpark Müngersdorf 6, 50933 Cologne, Germany

Center for Health and Integrative Physiology in Space, German Sport University Cologne, Am Sportpark Müngersdorf 6, 50933 Cologne, Germany
e-mail: v.abeln@dshs-koeln.de; T.Vogt@dshs-koeln.de

S. Schneider
Institute for Movement and Neurosciences, German Sport University Cologne, Am Sportpark Müngersdorf 6, 50933 Cologne, Germany

Center for Health and Integrative Physiology in Space, German Sport University Cologne, Am Sportpark Müngersdorf 6, 50933 Cologne, Germany

Faculty for Science, Health, Education and Engineering, University of the Sunshine Coast, Maroochydore DC, QLD 4558, Australia
e-mail: Schneider@dshs-koeln.de

S. Schneider (ed.), *Exercise in Space*, SpringerBriefs in Space Life Sciences,
DOI 10.1007/978-3-319-29571-8_5

5.1 Introduction

Space-related research of the past decades provides significant evidence that living and working in Space is bound to severe stressors for the human body and mind. These Space-induced stressors are mainly characterised by gravitational changes, severity of climate, sensorial deprivation, isolation and confinement affecting physical and mental health (Fowler and Manzey 2000; Manzey 2000). Although maintaining the integrity of physiological systems for the prevention of physical fitness is crucial for a safe return to Earth and even more for accelerating the progress of rehabilitation and recovery when returning to Earth, a deficient state of mental health may jeopardise mission safety and mission success during human Space explorations.

The issue of exercise-induced influences on physical health in Space is been extensively discussed in other chapters of this book. This chapter is dedicated to describe the psychophysiological effects of exercise in Space. The aim is to show that exercise in Space seems to have a more extensive effect than just staying physically fit. Exercise in Space might also help to improve mood, enhance neurocognitive function and therefore increase crew performance and mission safety and success.

In theory, life in Space may be considered as an analogue to a sedentary lifestyle on Earth. This reasoning is based on both physical and mental deconditioning throughout the lifespan that are caused by a sedentary lifestyle. Space-living conditions, especially weightlessness and reduced physical activity, contribute to accelerated deconditioning of physical and mental state. Thus, Space habitation might be considered as time-lapse of a sedentary lifestyle and ageing (Fig. 5.1). Accordingly, the ideas and content of this chapter may also contribute to present debates on active living and its relevance for socioeconomic and health-political decisions in the near and far future.

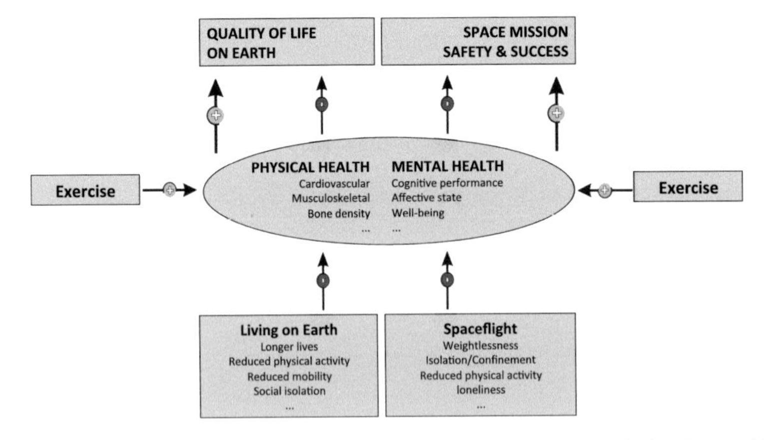

Fig. 5.1 The influence of living conditions on Earth and in Space on physical and mental health and life quality and Space mission success and safety as well as the role of exercise in this process

5.2 Physical Fitness in Space

The adaptation or de-adaptation of physiological systems such as the musculoskeletal or cardiovascular system under missing gravity conditions is an important field of research. Within the last four decades, it has provided us not only with innovative but also pioneering information, which are nowadays widely used inter alia in the rehabilitation of patients suffering from the negative effects of immobilisation, e.g. after surgery.

Everyone agrees that an adequate exercise program in Space prevents from physiological deconditioning of muscles, bones as well as the cardiovascular system. This helps astronauts and cosmonauts after long duration spaceflight to quickly readapt to Earth gravity. Although it has been shown that a regular exercise program will not completely avert deconditioning of the musculoskeletal system (Gopalakrishnan et al. 2010), the decremental effects of living under weightlessness conditions might be decelerated by a regular exercise routine. But while the recovery from physical deconditioning does not become important before having returned to Earth, mental impairments contain life- and mission-threatening risks for the ongoing mission in Space. Accordingly, it seems to be of even higher importance to find diagnostic and preventive tools for the prevention of mental impairments.

5.3 Mental Fitness in Space

The World Health Organization (WHO) defines mental health as a state of subjective well-being as a construct of biological, psychological and social factors (WHO 2001). There are a number of factors related to mental impairment such as stress, depression, anxiety, learning disabilities and mood disorders, which all impact the ability to function in everyday life. Mental impairments therefore include affective as well as cognitive changes. Everyday stressors, such as a high workload and socioeconomic demands, limited physical activity and social contacts, are known to increase the incidence of mental impairments in our general population.

Mental impairment is not only a concern for people living on Earth. People living in Space will increasingly face this problem because of the intention to extend Space mission duration and range, e.g. a journey to Mars (Slack et al. 2015). With an increasing duration of Space missions, the hostile environment, the limitations in the Space habitat, the tremendous mission workload, and last but not least the limited social interactions on board have been identified as main stressors. A spaceflight combines the risk factors for mental health, including biological (e.g. weightlessness, radiation, missing sunlight and day-night cycle, stress), psychological (e.g. mission goals, duration, workload, sensory and sleep deprivation) and social factors (e.g. crew dynamics, heterogeneous crews, loneliness) (Sandal et al. 2006; Rosnet et al. 2000). Thus, mental impairments are likely to occur.

5.3.1 Cognitive Changes in Space

Twenty years ago, the International Space Agencies (ESA, NASA) have identified cognitive performances to be of utterly importance for mission success and mission safety. Therefore, aviation and Space science extensively focussed on topics that addressed cognitive performance. Based on terrestrial research that provides evidence for an impact of stress on cognitive performance, similar stress-related deficits were expected for cognitive performance in Space (Albery 1989; Lieberman et al. 2002; Hockey 1997). However, Space and Space analogue studies showed divergent results. While evidence for cognitive impairments concerning psychomotor speed (Berger et al. 1997; Bock et al. 2001; Heuer et al. 2003), internal timekeeping (Semjen et al. 1998), attentional processes (Pattyn et al. 2005) and central management of concurrent tasks (Manzey and Lorenz 1999) exists, other studies fail to identify cognitive deficits (Manzey and Lorenz 1998; Eddy et al. 1998; Benke et al. 1993; Abeln et al. 2015; Schneider et al. 2012). However, there are reports of disturbed decision-making and judging in Space, which may have fatal consequences (Palinkas 1990; Wood et al. 2000). Further, a number of unreported behavioural health problems that refer to the confinement of astronauts and/or study participants are assumed. Consequently, it supports the notion that changes of cognitive performance should not be underestimated.

5.3.2 Affective Changes in Space and Analogue Environments

So far, emotional data has rarely been collected during spaceflight or is of minor validity due to limited number of subjects. The majority of assumptions are based on investigations in analogue environments like polar stations at Antarctica (see Picture 5.1). In polar studies substantial sleep disturbances and associated decreases

Picture 5.1 Impressions of isolation studies. Picture *left* shows a subject exercising during the Mars500 study on a self-powered treadmill. Picture *right* shows a participant at Concordia Station at Antarctica (*middle*) during an exercise session on a bicycle ergometer

of subjective well-being and alertness have been observed during the periods of missing sunlight (Bhargava et al. 2000; Natani et al. 1970; Palinkas et al. 1995a, b, 1996; Palinkas and Johnson 1990; Ikegawa et al. 1998). In addition, depressive and psychosomatic symptoms, crew conflicts and declined working performance have been shown (Suedfeld 2005; Kanas and Manzey 2003). During spaceflights, psychological reactions, such as decreased attention, vigour, motivation, appetite as well as higher emotional and perceptual sensitivity, fatigue, psychosomatic symptoms and interpersonal conflicts, were reported (Kanas et al. 2001a, b; Kanas 1991; Suedfeld 2005; Kanas and Manzey 2003).

Similar to the variations in cognitive performance within different crews, the existing results of long-term isolation on affective states during spaceflight and analogue conditions are inconsistent. While some crews show affective declines around mission midpoint (Schneider et al. 2010) or within the third quarter (Palinkas and Suedfeld 2008), some show declining emotions over mission duration (Abeln et al. 2015) and others no emotional declines at all or even positive outcomes (Sandal et al. 2006; Palinkas and Suedfeld 2008; Wood et al. 2000; Palinkas et al. 1995b; Kanas 1998). Interestingly, affective state and individual performance seem to be correlating (Manzey et al. 1998; Lorenz et al. 1996). This correlation underlines the request to consider and counteract both affective state and individual performance.

5.3.3 Underlying Neurophysiological Mechanisms

Although it has been shown that the multi-stressor environment of a spaceflight has a negative impact on mental health (Fowler and Manzey 2000; Kanas 1998), the underlying neurophysiological processes remain widely unclear. This is mainly due to a limited number of Space missions and, thus, possibilities to collect respective data as well as a lack of imaging technologies to record respective data in Space (Schneider et al. 2012). Several stressors that occur during a Space mission, such as sleep disorders, the absence of sunlight and radiation or altered physical activity, have multifactorial consequences and initiate a lot of interactions within the human body, which challenge the scientists to create a comprehensive picture. Nevertheless, the following neurophysiological mechanisms aim to integrate previous findings and possible approaches for future investigations.

5.3.3.1 Haemodynamic Changes

Fluid/blood shifts towards the upper part of the body and the brain in weightlessness were suggested to cause changes in cognitive performance. Head-down tilt studies aimed to simulate the increase of cerebral blood volume in order to explore the underlying mechanisms of cognitive impairments in Space. However, a connection between blood volume shifts and cognitive functions was not found (Pavy Le-Traon et al. 1994; Shehab et al. 1998). During parabolic flights, blood volume shifts towards the brain during microgravity and specific haemodynamic responses to hypo- and hypergravity were detected via functional near-infrared spectroscopy (Schneider et al. 2013a). No correlation of cerebral haemodynamic shifts and electrocortical activity was shown during artificial gravity exposure using a short-arm human centrifuge (Smith et al. 2013). Interestingly, preliminary findings even demonstrate higher cognitive performances in relation to increased amounts of blood and, thus, oxygen (Wollseiffen et al. 2016). Accordingly, blood volume shifts in weightlessness can be excluded to be the reason for cognitive decline in Space.

5.3.3.2 Endocrinological Changes

Recent research reports on significant effects of the acute stressor weightlessness on endocrinological regulations. Based on blood sampling during rather unique and stressful parabolic flight manoeuvres, the levels of stress hormones prolactin, cortisol and ACTH (Schneider et al. 2007), as well as endocannabinoids (Strewe et al. 2012), increased. Stress hormone concentration was also elevated during a 6-month exposure of weightlessness on the ISS (Strewe et al. 2012) as well as during long-term confinement in analogue environments (Strollo et al. 2014; Yi et al. 2015; Jacubowski et al. 2015). Without being aware of the level of impact, stress hormones are considered to play a role for the response and adaptation in physiological and psychological systems. Further investigations are necessary to more clearly determine the importance of endocrinological changes and their impact on cognitive and affective performance. Endocrinological screenings prior to as well as during Space travels might be a useful tool for the prevention and acute threatening of illness and subject recruitment.

5.3.3.3 Electrocortical Changes

Due to methodological restrictions, studies examining electrocortical changes in Space or analogue environments are rare (Schneider et al. 2012). Some first attempts report of stress-related changes expressed by increased electroencephalographic (EEG) beta activity or higher functional asymmetry between hemispheres during parabolic flights (De Metz et al. 1994; Pletser and Quadens 2003). Advanced technologies using electrocortical recordings combined with electromagnetic tomography and near-infrared spectroscopy allowed for cortical imaging during weightlessness. Combined research programs including parabolic flight, lower body negative pressure, human centrifuge manoeuvres, as well as supine, seated and head-down tilt postures, aimed to address neurophysiological attributes related to changed gravity conditions (i.e. micro- and hypergravity). A number of studies lead to the conclusion that electrocortical activation does not relate to central haemodynamic shifts but might reflect neuro-affective and neurocognitive adaptations (Schneider et al. 2008a, c, 2009c, 2013a; Smith et al. 2013; Dern et al. 2014; Hunt et al. 2006; Kurihara et al. 2003; Yasumasa et al. 2002). This is particularly supported by increased activity within the frontal lobe, which is associated with emotional and cognitive processes (Schneider et al. 2008b, 2009c; Smith et al. 2013).

The Mars105-day isolation trial revealed decreasing brain activity (alpha and beta frequencies) in the course of the first two thirds of the isolation period and accompanying affective declines (perceived physiological and motivation state, psychological strain) as well as a recovery prior to the end (Schneider et al. 2010). In addition, the Mars520-day isolation trial revealed decreasing electrocortical activity and increasing cortisol levels (Jacubowski et al. 2015). Others showed changes of electrocortical activity related to cognitive performance in isolation (Lorenz et al. 1996; Schneider et al. 2013b). Thus, isolation/confinement seems to cause an irritation of central nervous state mediating or modulated by psychophysiological stress.

So far little attention has been given to long-term investigations of brain cortical activity in Space (Cheron et al. 2006), which makes it difficult to draw any manifest conclusion. The implementation of new integrative, easily applicable and usable neuroimaging technologies (EEG, fNIRS) in the International Space Station (ISS) is planned, which are promising tools contributing to new insights of the effect of long-term exposure to weightlessness and its circumstances on neurophysiological and neuropsychological state. Recording of electrocortical activity offers to explore the process of neurocognitive and neuro-affective changes in Space and analogue environments. Also, this might open new possibilities for crew screening and observations during missions.

To conclude, (1) Space science in terms of mental health is still in its infancy, and (2) comparability of every single space and Space analogue mission is difficult and the progress, particularly of psychological state, is often unpredictable. Even minor events or changes may result in decreasing motivation, mood and/or compliance. Subsequently, accidents, emergencies, crew conflicts, health problems or other events may happen. As a consequence of this, there is a strong need for countermeasures preventing mental impairment.

5.4 Countermeasures for Neurocognitive and Neuro-affective Changes in Space

In the past decades, Space research mainly focused on investigating the effects of stressors of spaceflight, especially weightlessness towards physical health. Thus, the time seems right to initiate the next step and to concentrate on countermeasures to prevent Space-related stressors. Existing knowledge offers to be applied to actually support astronauts, cosmonauts and taikonauts in flying and living in

Space and to support mission success and mission safety. This is even more important, when taking longer lasting Space missions of the future into account.

Attempts in terms of proper planning and preparation, such as, e.g. accurate and anticipatory calculations of energy, power, consumables, and nutrition as well as emergency strategies and practice trials, an elaborated selection of candidates and consideration of individual needs and preferences (Flynn 2005; Slack et al. 2015; Sandal et al. 2006; Kanas 1998), certainly help to secure mission safety and mission success in order to prevent many of the risks that spaceflights may implicate. Moreover, crew selection based on geno- and phenotypes (Van Dongen et al. 2004; Kuna et al. 2012), biomarkers (Czeisler 2011), or managing strategies, e.g. of sleep-wake regulation (Goel and Dinges 2012), has been discussed. However, spaceflight history taught that unpredictable events and changes of a subjective state of physical and mental health require more independent and holistic countermeasures.

5.4.1 *Artificial Gravity*

One approach that has recently been discussed as a countermeasure for Space-related impairments within the cardiovascular, musculoskeletal and central nervous system is artificial gravity provided by a short-arm human centrifuge (SAHC). The first results on the effects of an ESA-owned SAHC indicate that exposure to artificial gravity results in a physical activation that is similar to standing (Goswami et al. 2015a). Neglecting the possible occurrence of motion sickness and syncope, one might think about the benefit of "passive exercise", at least at moderate G-levels of 1–1.5G. Instead of time-consuming active exercise, passive exercise, i.e. artificial gravity, in a supine or seated position could be used for relaxation or individual leisure time (e.g. reading a book, viewing a DVD). One might even think about being centrifuged while sleeping. But although there are physiological effects suggesting implications for orthostatic tolerance training (Goswami et al. 2015b) and the musculoskeletal and cardiovascular system (Goswami et al. 2015a; Iwasaki et al. 2005; Iwase et al. 2002; Symons et al. 2009; Caiozzo et al. 2009), no benefit for affective state and mental performance could be identified (Dern et al. 2014; Vogt et al. 2014). In contrast, there is evidence that artificial gravity causes additional mental stress, which is underlined by increased electrocortical activity and deoxygenation in the frontal cortex (Smith et al. 2013). Therefore artificial gravity maybe questioned to sufficiently serve as an exercise-like countermeasure for Space-related deteriorations.

5.4.2 Physical Activity/Exercise

While the benefits of exercise on cardiovascular responses as well as musculoskeletal functionality are widely accepted, exercise seems to provide benefits not only for physical fitness but also for mental health (Martinsen 2000). This includes the enhancement of cognitive performance (Biddle and Asare 2011; Blumenthal et al. 1991; Lojovich 2010) as well as an improvement of affective state (Wolff et al. 2011; Vogt et al. 2012; Ekkekakis and Acevedo 2006; Parfitt et al. 2006; Schneider et al. 2009a). It is only within the past few years that earthbound research on exercise programs and their influence on mental health indicates a substantial involvement of exercise-induced central neuronal adaptations (Ekkekakis and Acevedo 2006; Schneider et al. 2009b). In particular, frontal brain regions that are associated to play an important role in information as well as emotional processing seem to be positively affected by exercise (Brümmer et al. 2011; Schneider et al. 2009b). These modifications are supposed to act as a multifunctional generator for the adaptation of mood, vigilance and cognitive performance. Accordingly, medical interests on embedding such beneficial exercise programs, to enhance neurocognitive functioning as well as to improve neuro-affective processing, is increasing. The aim is to counteract emerging pathological challenges of our ageing society, to reduce consequences of sedentary lifestyles such as neurological degenerations.

Within Space research, exercise is constantly used as a countermeasure against cardiovascular and musculoskeletal deconditioning. So far, the implication of exercise for mental health has been neglected also due to the relatively young and deficient state of literature in this respect. However, applying exercise not only for physical but also for mental health has gained interests within the past years as a consequence of the abovementioned positive outcomes of exercise interventions on Earth and recent Space-related investigations (Abeln et al. 2015; Basner et al. 2013; Schneider et al. 2010, 2013b). Basner and colleagues just recently reported a correlation between decreased physical activity levels, sleep quality and quantity and vigilance during the Mars520 project. Within the Mars105-day trial, an enhancing effect of a short bout of exercise on brain cortical activity and perceived physical, psychological and motivational state could be shown [see Table 5.1 and Schneider et al. (2010)]. Table 5.1 provides comparable findings and a decline of mood within an isolated and inactive cohort of crewmembers during overwintering in Antarctica, whereas regularly active crewmembers did not show any impairment (Abeln et al. 2015).

Table 5.1 The effect of isolation on electrocortical activation and affective state

Group	Parameter	Iso-1		Iso-2		Iso-3		Iso-4	
		Inactive	Active	Inactive	Active	Inactive	Active	Inactive	Active
Concordia	EEG alpha	1.21	2.38	2.01	2.18	1.84	1.97	2.00	1.99
Mars105	EEG alpha	2.17	2.43	1.90	2.18	1.81	2.23	2.40	2.36
Concordia	EEG beta	1.02	1.18	1.29	1.23	1.28	0.88	1.08	1.04
Mars105	EEG beta	1.12	1.61	0.70	1.23	0.65	0.98	1.18	1.33
Concordia	PEPS	4.22	3.89	3.72	4.22	3.33	4.06	3.30	3.83
Mars105	PEPS	4.14	4.14	4.14	4.14	3.98	3.98	4.14	4.14
Concordia	PSYCHO	4.00	3.72	3.28	3.84	3.03	3.53	2.78	3.72
Mars105	PSYCHO	3.81	4.10	3.48	3.73	3.48	3.88	3.65	4.02
Concordia	MOT	4.02	3.75	3.47	3.82	3.38	3.81	3.25	4.06
Mars105	MOT	3.77	3.92	3.56	3.77	3.27	3.60	3.71	3.81

The table shows results of two isolation studies: (1) overwintering at Concordia Station in Antarctica ($n = 8$), shown are the first four measurements of 6-week intervals during 9 months isolation for those who exercised on a regular basis (active) and those who did not or not regularly exercise (inactive); (2) 105-day isolation mission of Mars500 project ($n = 6$), shown are four measurements of 4-week intervals during 105-day isolation pre- (inactive) and postexercise (active). For both isolation studies, electroencephalographic (EEG) alpha and beta frequency as well as perceived physical (PEPS) and motivational state (MOT), as well as perceived psychological strain (PSYCHO), please refer to Abeln et al. (2015) and Schneider et al. (2010)

Exercise certainly is one promising holistic approach against Space- and isolation-related impairments but also for related impairments on Earth. In this regard, we would like to point out that people in Space and on Earth will profit from investigations in both settings and that research should aim to be complementary.

5.5 Exercise in Space: Recommendations for the Future

Which exercise programs are necessary for the benefit of mental health, in particular when facing the restrictive circumstances as present during spaceflights?

In consideration of the above-stated incidence of physical and mental impairments in Space as well as the potential benefit of exercise to serve as a holistic countermeasure against both, there is an essential need for national and international Space agencies to define the aims of future "exercise in Space" programs. If decided to follow a holistic approach of exercise, it is of utterly importance to understand the beneficial impact of exercise programs also on neurocognitive functioning and neuro-affective processing as markers of cognitive performance, mental health and well-being. These specific exercise programs for mental health need to take into account several outcomes of earthbound exercise study.

First, it has to be mentioned that exercise mode, duration and intensity seems to matter. Especially, running exercise causes a higher state of cortical relaxation within frontal brain regions, which are associated with higher executive functions (Brümmer et al. 2011; Schneider et al. 2009a, b). Furthermore, the consideration of individual exercise preferences and freely chosen instead of prescribed exercise protocols seems to have an important impact (Ekkekakis 2009). Exercise was shown to reveal greater positive affective responses when the intensity was below the individual ventilatory threshold and when the intensity was self-selected, even if the same intensity was prescribed. Moreover, the review by Ekkekakis (2009) also discovered that people on average automatically chose an intensity lying within the recommended "healthy range". There is also Space-related evidence for this individual preference hypothesis: Schneider et al. (2013b) were able to show that running exercise resulted in frontal cortex deactivation (relaxation) and higher cognitive performance during long-term isolation. Interestingly, the crew indicated running to be the most preferred exercise.

Accordingly and with respect to mental health, exercise does not need to be prescribed, but on the contrary seems to be even more effective when self-selected and controlled. Motivation, commitment and compliance for exercise, which are obviously higher when individual preferences are appreciated, are aspects that should not be underestimated when we aim for regular exercise to obtain physical and mental health at longer-duration Space missions. It has to be investigated whether the effect of self-controlled exercise in Space provokes effects comparable to the presently prescribed exercise protocols. Because, astronauts/cosmonauts and taikonauts currently spend high amounts of time exercising (about 2–2.5 h daily) without being able to completely counteract physiological deconditioning

(Gopalakrishnan et al. 2010), it seems even reasonable to replace some of the prescribed exercise sessions with arbitrary exercise sessions.

This individualisation of exercise programs might also be of relevance for earthbound therapies using exercise programs to avoid sedentary lifestyles with all its negative effects including physical and mental degeneration. Technological developments that are supportive to the individual's motivation to exercise—both on Earth and in Space—might be useful, such as the implementation of virtual realities, allowing for jogging or cycling in familiar environments like a forest home track.

References

Abeln V, Macdonald-Nethercott E, Piacentini MF, Meeusen R, Kleinert J, Strueder HK, Schneider S (2015) Exercise in isolation – a countermeasure for electrocortical, mental and cognitive impairments. PLoS One 10:e0126356

Albery WB (1989) The effect of sustained acceleration and noise on workload in human operators. Aviat Space Environ Med 60:943–948

Basner M, Dinges DF, Mollicone D, Ecker A, Jones CW, Hyder EC, Di Antonio A, Savelev I, Kan K, Goel N, Morukov BV, Sutton JP (2013) Mars 520-d mission simulation reveals protracted crew hypokinesis and alterations of sleep duration and timing. Proc Natl Acad Sci USA 110:2635–2640

Benke T, Koserenko O, Watson NV, Gerstenbrand F (1993) Space and cognition: the measurement of behavioral functions during a 6-day space mission. Aviat Space Environ Med 64:376–379

Berger M, Mescheriakov S, Molokanova E, Lechner-Steinleitner S, Seguer N, Kozlovskaya I (1997) Pointing arm movements in short- and long-term spaceflights. Aviat Space Environ Med 68:781–787

Bhargava R, Mukerji S, Sachdeva U (2000) Psychological impact of the Antarctic winter on Indian expeditioners. Environ Behav 32:111–127

Biddle SJ, Asare M (2011) Physical activity and mental health in children and adolescents: a review of reviews. Br J Sports Med 45:886–895

Blumenthal JA, Emery CF, Madden DJ, Schniebolk S, Walsh-Riddle M, George LK, Mckee DC, Higginbotham MB, Cobb FR, Coleman RE (1991) Long-term effects of exercise on psychological functioning in older men and women. J Gerontol 46:352–361

Bock O, Fowler B, Comfort D (2001) Human sensorimotor coordination during spaceflight: an analysis of pointing and tracking responses during the "Neurolab" space shuttle mission. Aviat Space Environ Med 72:877–883

Brümmer V, Schneider S, Abel T, Vogt T, Strüder HK (2011) Brain cortical activity is influenced by exercise mode and intensity. Med Sci Sports Exerc 43:1863–1872

Caiozzo VJ, Haddad F, Lee S, Baker M, Paloski W, Baldwin KM (2009) Artificial gravity as a countermeasure to microgravity: a pilot study examining the effects on knee extensor and plantar flexor muscle groups. J Appl Physiol (1985) 107:39–46

Cheron G, Leroy A, De Saedeleer C, Bengoetxea A, Lipshits M, Cebolla A, Servais L, Dan B, Berthoz A, Mcintyre J (2006) Effect of gravity on human spontaneous 10-Hz electroencephalographic oscillations during the arrest reaction. Brain Res 1121:104–116

Czeisler CA (2011) Impact of sleepiness and sleep deficiency on public health—utility of biomarkers. J Clin Sleep Med 7:S6–S8

De Metz K, Quadens O, De Graeve M (1994) Quantified EEG in different G situations. Acta Astronaut 32:151–157

Dern S, Vogt T, Abeln V, Struder HK, Schneider S (2014) Psychophysiological responses of artificial gravity exposure to humans. Eur J Appl Physiol 114:2061–2071
Eddy DR, Schifflett SG, Schlegel RE, Shehab RL (1998) Cognitive performance aboard the life and microgravity spacelab. Acta Astronaut 43:193–210
Ekkekakis P, Acevedo EO (2006) Affective response to acute exercise: toward a psychobiological dose-response model. In: Acevedo EO, Ekkekakis P (eds) Psychobiology of physical activity. Human Kinetics, Champaign
Ekkekakis P (2009) Let them roam free? Physiological and psychological evidence for the potential of self-selected exercise intensity in public health. Sports Med 39(10):857–888
Flynn CF (2005) An operational approach to long-duration mission behavioral health and performance factors. Aviat Space Environ Med 76:B42–B51
Fowler B, Manzey D (2000) Summary of research issues in monitoring of mental and perceptual-motor performance and stress in space. Aviat Space Environ Med 71:A76–A77
Goel N, Dinges DF (2012) Predicting risk in space: genetic markers for differential vulnerability to sleep restriction. Acta Astronaut 77:207–213
Gopalakrishnan R, Genc KO, Rice AJ, Lee SM, Evans HJ, Maender CC, Ilaslan H, Cavanagh PR (2010) Muscle volume, strength, endurance, and exercise loads during 6-month missions in space. Aviat Space Environ Med 81:91–102
Goswami N, Bruner M, Xu D, Bareille MP, Beck A, Hinghofer-Szalkay H, Blaber AP (2015a) Short-arm human centrifugation with 0.4g at eye and 0.75g at heart level provides similar cerebrovascular and cardiovascular responses to standing. Eur J Appl Physiol 115:1569–1575
Goswami N, Evans J, Schneider S, Von der Wiesche M, Mulder E, Rossler A, Hinghofer-Szalkay H, Blaber AP (2015b) Effects of individualized centrifugation training on orthostatic tolerance in men and women. PLoS One 10:e0125780
Heuer H, Manzey D, Lorenz B, Sangals J (2003) Impairments of manual tracking performance during spaceflight are associated with specific effects of microgravity on visuomotor transformations. Ergonomics 46:920–934
Hockey GR (1997) Compensatory control in the regulation of human performance under stress and high workload; a cognitive-energetical framework. Biol Psychol 45:73–93
Hunt K, Tachtsidis I, Bleasdale-Barr K, Elwell C, Mathias C, Smith M (2006) Changes in cerebral oxygenation and haemodynamics during postural blood pressure changes in patients with autonomic failure. Physiol Meas 27:777–785
Ikegawa M, Kimura M, Makita K, Itokawa Y (1998) Psychological studies of a Japanese winter-over group at Asuka Station, Antarctica. Aviat Space Environ Med 69:452–460
Iwasaki K, Shiozawa T, Kamiya A, Michikami D, Hirayanagi K, Yajima K, IWASE S, Mano T (2005) Hypergravity exercise against bed rest induced changes in cardiac autonomic control. Eur J Appl Physiol 94:285–291
Iwase S, Fu Q, Narita K, Morimoto E, Takada H, Mano T (2002) Effects of graded load of artificial gravity on cardiovascular functions in humans. Environ Med 46:29–32
Jacubowski A, Abeln V, Vogt T, Yi B, Chouker A, Fomina E, Struder HK, Schneider S (2015) The impact of long-term confinement and exercise on central and peripheral stress markers. Physiol Behav 152:106–111
Kanas N (1991) Psychosocial support for cosmonauts. Aviat Space Environ Med 62:353–355
Kanas N (1998) Psychiatric issues affecting long duration space missions. Aviat Space Environ Med 69:1211–1216
Kanas N, Manzey D (2003) Space psychology and psychiatry. Kluwer, Dodrecht
Kanas N, Salnitskiy V, Grund EM, Weiss DS, Gushin V, Bostrom A, Kozerenko O, Sled A, Marmar CR (2001a) Psychosocial issues in space: results from Shuttle/Mir. Gravit Space Biol Bull 14:35–45
Kanas N, Salnitskiy V, Grund EM, Weiss DS, Gushin V, Kozerenko O, Sled A, Marmar CR (2001b) Human interactions during Shuttle/Mir space missions. Acta Astronaut 48:777–784

Kuna ST, Maislin G, Pack FM, Staley B, Hachadoorian R, Coccaro EF, Pack AI (2012) Heritability of performance deficit accumulation during acute sleep deprivation in twins. Sleep 35:1223–1233

Kurihara K, Kikukawa A, Kobayashi A (2003) Cerebral oxygenation monitor during head-up and -down tilt using near-infrared spatially resolved spectroscopy. Clin Physiol Funct Imaging 23:177–181

Lieberman HR, Tharion WJ, Shukitt-Hale B, Speckman KL, Tulley R (2002) Effects of caffeine, sleep loss, and stress on cognitive performance and mood during U.S. Navy SEAL training. Sea-Air-Land. Psychopharmacology (Berl) 164:250–261

Lojovich JM (2010) The relationship between aerobic exercise and cognition: is movement medicinal? J Head Trauma Rehabil 25:184–192

Lorenz B, Lorenz J, Manzey D (1996) Performance and brain electrical activity during prolonged confinement. Adv Space Biol Med 5:157–181

Manzey D (2000) Monitoring of mental performance during spaceflight. Aviat Space Environ Med 71:A69–A75

Manzey D, Lorenz B (1998) Mental performance during short-term and long-term spaceflight. Brain Res Rev 28:215–221

Manzey D, Lorenz B (1999) Human performance during spaceflight. Hum Perf Extrem Environ 4:8–13

Manzey D, Lorenz B, Poljakov V (1998) Mental performance in extreme environments: results from a performance monitoring study during a 438-day spaceflight. Ergonomics 41:537–559

Martinsen EW (2000) Physical activity for mental health. Tidsskr Nor Laegeforen 120:3054–3056

Natani K, Shurley JT, Pierce CM, Brooks RE (1970) Long-term changes in sleep patterns in men on the South Polar Plateau. Arch Intern Med 125:655–659

Palinkas LA (1990) Psychosocial effects of adjustment in Antarctica: lessons for long-duration spaceflight. J Spacecr Rocket 27:471–477

Palinkas LA, Johnson JC (1990) Social relations and individual performance of winter-over personnel at McMurdo Station. Antarct J US 25:238–240

Palinkas LA, Suedfeld P (2008) Psychological effects of polar expeditions. Lancet 371:153–163

Palinkas LA, Cravalho M, Browner D (1995a) Seasonal variation of depressive symptoms in Antarctica. Acta Psychiatr Scand 91:423–429

Palinkas LA, Suedfeld P, Steel GD (1995b) Psychological functioning among members of a small polar expedition. Aviat Space Environ Med 66:943–950

Palinkas LA, Houseal M, Rosenthal NE (1996) Subsyndromal seasonal affective disorder in Antarctica. J Nerv Ment Dis 184:530–534

Parfitt G, Rose EA, Burgess WM (2006) The psychological and physiological responses of sedentary individuals to prescribed and preferred intensity exercise. Br J Health Psychol 11:39–53

Pattyn N, Migeotte P, Demaesselaer W, Kolinsky R, Morais J, Zizi M (2005) Investigating human cognitive performance during spaceflight. J Gravit Physiol 12:2

Pavy Le-Traon A, Rous De Feneyrols A, Cornac A, Abdeseelam R, N'uygen D, Lazerges M, Guell A, Bes A (1994) Psychomotor performance during a 28 day head-down tilt with and without lower body negative pressure. Acta Astronaut 32:319–330

Pletser V, Quadens O (2003) Degraded EEG response of the human brain in function of gravity levels by the method of chaotic attractor. Acta Astronaut 52:581–589

Rosnet E, Le Scanff C, Sagal MS (2000) How self-image and personality influence performance in an isolated environment. Environ Behav 32:18–31

Sandal G, Leon G, Palinkas L (2006) Human challenges in polar and space environments. Rev Environ Sci Biotechnol 5:281–296

Schneider S, Brümmer V, Gobel S, Carnahan H, Dubrowski A, Strüder HK (2007) Parabolic flight experience is related to increased release of stress hormones. Eur J Appl Physiol 100:301–308

Schneider S, Askew CD, Brümmer V, Kleinert J, Guardiera S, Abel T, Strüder HK (2008a) The effect of parabolic flight on perceived physical, motivational and psychological state in men and women: correlation with neuroendocrine stress parameters and electrocortical activity. Stress 12:336–349

Schneider S, Brümmer V, Carnahan H, Dubrowski A, Askew CD, Strüder HK (2008b) What happens to the brain in weightlessness? A first approach by EEG tomography. Neuroimage 42:1316–1323

Schneider S, Guardiera S, Kleinert J, Steinbacher A, Abel T, Carnahan H, Struder HK (2008c) Centrifugal acceleration to 3Gz is related to increased release of stress hormones and decreased mood in men and women. Stress 11:339–347

Schneider S, Askew CD, Diehl J, Mierau A, Kleinert J, Abel T, Carnahan H, Strüder HK (2009a) EEG activity and mood in health orientated runners after different exercise intensities. Physiol Behav 96:709–716

Schneider S, Brümmer V, Abel T, Askew CD, Strüder HK (2009b) Changes in brain cortical activity measured by EEG are related to individual exercise preferences. Physiol Behav 98:447–452

Schneider S, Guardiera S, Abel T, Carnahan H, Struder HK (2009c) Artificial gravity results in changes in frontal lobe activity measured by EEG tomography. Brain Res 1285:119–126

Schneider S, Brümmer V, Carnahan H, Kleinert J, Piacentini MF, Meeusen R, Strüder HK (2010) Exercise as a countermeasure to psycho-physiological deconditioning during long-term confinement. Behav Brain Res 211:208–214

Schneider S, Bubeev JA, Chouker A, Morukov BV, Johannes B, Strueder HK (2012) Imaging of neuro-cognitive performance in extreme environments – a (p)review. Planet Space Sci 74:7

Schneider S, Abeln V, Askew CD, Vogt T, Hoffmann U, Denise P, Struder HK (2013a) Changes in cerebral oxygenation during parabolic flight. Eur J Appl Physiol 113:1617–1623

Schneider S, Abeln V, Popova J, Fomina E, Jacubowski A, Meeusen R, Struder HK (2013b) The influence of exercise on prefrontal cortex activity and cognitive performance during a simulated space flight to Mars (MARS500). Behav Brain Res 236:1–7

Semjen A, Leone G, Lipshits M (1998) Temporal control and motor control: two functional modules which may be influenced differently under microgravity. Hum Mov Sci 17:77–93

Shehab RL, Schlegel RE, Schiflett SG, Eddy DR (1998) The NASA performance assessment workstation: cognitive performance during head-down bed rest. Acta Astronaut 43:223–233

Slack KJ, Schneiderman JS, Leveton LB, Whitmire AM, Picano JJ (2015) Evidence report: risk of adverse cognitive or behavioral conditions and psychiatric disorders. In: Human Research Program, BHAP, National Aeronautics And Space Administration (ed) http://humanresearchroadmap.nasa.gov/evidence/reports/BMED.pdf. Houston

Smith C, Goswami N, Robinson R, Von der Wiesche M, Schneider S (2013) The relationship between brain cortical activity and brain oxygenation in the prefrontal cortex during hypergravity exposure. J Appl Physiol (1985) 114:905–910

Strewe C, Feuerecker M, Nichiporuk I, Kaufmann I, Hauer D, Morukov B, Schelling G, Chouker A (2012) Effects of parabolic flight and spaceflight on the endocannabinoid system in humans. Rev Neurosci 23:673–680

Strollo F, Vassilieva G, Ruscica M, Masini M, Santucci D, Borgia L, Magni P, Celotti F, Nikiporuc I (2014) Changes in stress hormones and metabolism during a 105-day simulated Mars mission. Aviat Space Environ Med 85:793–797

Suedfeld P (2005) Invulnerability, coping, salutogenesis, integration: four phases of space psychology. Aviat Space Environ Med 76:B61–B66

Symons TB, Sheffield-Moore M, Chinkes DL, Ferrando AA, Paddon-Jones D (2009) Artificial gravity maintains skeletal muscle protein synthesis during 21 days of simulated microgravity. J Appl Physiol (1985) 107:34–38

Van Dongen HP, Baynard MD, Maislin G, Dinges DF (2004) Systematic interindividual differences in neurobehavioral impairment from sleep loss: evidence of trait-like differential vulnerability. Sleep 27:423–433

Vogt T, Schneider S, Abeln V, Anneken V, Struder HK (2012) Exercise, mood and cognitive performance in intellectual disability – a neurophysiological approach. Behav Brain Res 226:473–480

Vogt T, Abeln V, Struder HK, Schneider S (2014) Artificial gravity exposure impairs exercise-related neurophysiological benefits. Physiol Behav 123:156–161

WHO (2001) The World health report: 2001: mental health: new understanding, new hope. WHO, Geneva

Wolff E et al (2011) Exercise and physical activity in mental disorders. Eur Arch Psychiatry Clin Neurosci 261(Suppl 2):S186–S191

Wollseiffen P et al (2016) Neuro-cognitive performance is enhanced during short periods of microgravity. Physiol Behav 155:9–16

Wood J, Hysong SJ, Lugg DJ, Harm DL (2000) Is it really so bad? A comparison of positive and negative experiences in Antarctic winter stations. Environ Behav 32:84–110

Yasumasa A, Inoue S, Tatebayashi K, Shiraishi Y, Kawai Y (2002) Effects of head-down tilt on cerebral blood flow in humans and rabbits. J Gravit Physiol 9:89–90

Yi B, Matzel S, Feuerecker M, Horl M, Ladinig C, Abeln V, Chouker A, Schneider S (2015) The impact of chronic stress burden of 520-d isolation and confinement on the physiological response to subsequent acute stress challenge. Behav Brain Res 281:111–115